ITCA Information Technology Certified Associate

SOFTWARE DEVELOPMENT FUNDAMENTALS

Study Guide

Part of the ITCA Stackable Certificate Program
www.isaca.org/itca

ISACA®

About ISACA

For more than 50 years, ISACA® (www.isaca.org) has advanced the best talent, expertise and learning in technology. ISACA equips individuals with knowledge, credentials, education and community to progress their careers and transform their organizations, and enables enterprises to train and build quality teams. Among those credentials, ISACA advances and validates business-critical skills and knowledge through the globally respected Certified Information Systems Auditor® (CISA®), Certified in Risk and Information Systems Control® (CRISC®), Certified Information Security Manager® (CISM®), Certified in the Governance of Enterprise IT® (CGEIT®) and Certified Data Privacy Solutions Engineer™ (CDPSE™) credentials. ISACA is a global professional association and learning organization that leverages the expertise of its 145,000 members who work in information security, governance, assurance, risk and privacy to drive innovation through technology. It has a presence in 188 countries, including more than 220 chapters worldwide.

Disclaimer

ISACA has designed and created *Software Development Fundamentals Study Guide* (the "Work") primarily as an educational resource for professionals. ISACA makes no claim that use of any of the Work will assure a successful outcome. The Work should not be considered inclusive of all proper information, procedures and tests or exclusive of other information, procedures and tests that are reasonably directed to obtaining the same results. In determining the propriety of any specific information, procedure or test, professionals should apply their own professional judgment to the specific circumstances presented by the particular systems or information technology environment.

Reservation of Rights

ISACA
1700 E. Golf Road, Suite 400
Schaumburg, IL 60173, USA
Phone: +1.847.660.5505
Fax: +1.847.253.1755
Contact us: https://support.isaca.org
Website: www.isaca.org

Provide Feedback: To comment, please visit https://support.isaca.org.
Participate in the ISACA Online Forums: https://engage.isaca.org/onlineforums

Twitter: http://twitter.com/ISACANews
LinkedIn: www.linkedin.com/company/isaca
Facebook: www.facebook.com/ISACAGlobal
Instagram: www.instagram.com/isacanews/

ISBN 978-1-60420-749-1
Software Development Fundamentals Study Guide
Printed in the United States of America

Acknowledgments

The *Software Development Fundamentals Study Guide* is the result of the collective efforts of many developers and volunteers. Special thanks go to:

Lead Developer

Sarit Williams, COBIT, CSM, CSTP, PMP, 360Adept.com, USA

Expert Reviewers

Christopher Agboile, CISA, CEH, MCSD, MSc Information Systems (Systems Security)
Matthew Altman, CISA, CRISC, CISM, CGEIT
Joyce Chua, CISA, CISM
Jacky Y.K. Ng, EngD, CISM, CDPSE, CEng, COBIT 5 Certified Assessor, CMgr, FCMI, MEIT
Dr. Y Rykalova, UMass Lowell, MA, USA
Marilize Van Schalkwykm, CISA, CIA

Page intentionally left blank

Table of Contents

Page intentionally left blank

LIST OF FIGURES

Chapter 1. Software Languages

Chapter 2. Programming

Chapter 3. Software Development Life Cycle

Page intentionally left blank

Introduction

Software development is the foundation of the technological improvements that benefit most people and enterprises today.

Software is interwoven into almost every aspect of our lives; for example, in:

- **Health**—Test machines, e.g., electrocardiogram (EKC), vitals machines, patient-record systems and fitness equipment
- **Finances**—Automated teller machines (ATMs) and online banking, e.g., the ability to send money electronically to someone miles away
- **Personal family and relationships**—Social software to keep in touch, take and send photographs, send a text message and communicate via phones that condense massive amounts of knowledge into a small, portable device

These are just a few small examples of where software is used. The world is increasingly dependent on the creation of software to solve everyday problems. Software development is the "set of computer science activities dedicated to the process of creating, designing, deploying and supporting software," which is the set of instructions (programs) that control a computer.[1] The development of the graphical user interface allowed people to easily interact with the underlying computer operating system software. For example, the ATM display screen is the user interface that makes it easy for bank customers to access their bank accounts.

The *Software Development Fundamentals Study Guide* introduces readers to software development, its history and evolution, and it equips them with basic essentials necessary to parlay this knowledge into the many domains of the software development field. This guide also helps to prepare individuals for the ISACA® Software Development Fundamentals certificate exam, one of the components of the ISACA Information Technology Certified Associate™ certification program.

The *Software Development Fundamentals Study Guide* covers three main areas:

- Software languages
- Programming
- Software development life cycle (SDLC) and management

Each chapter begins with learning objectives, and each chapter ends with a knowledge-check section.

1 IBM, "What is Software Development?," Software Development, www.ibm.com/topics/software-development

Page intentionally left blank

SOFTWARE DEVELOPMENT FUNDAMENTALS
An ISACA Certificate

Chapter 1:

Software Languages

Software Languages

Software Languages

1.1 Learning Objectives

After completing this chapter, learners will be able to:

1. Identify and explain basic software concepts and terms.
2. Describe software types and their uses.
3. Summarize the history of software (mainframe, personal computers and software-as-a-service eras).
4. Describe algorithms (problem analysis, development, modeling, etc.).
5. Differentiate between syntax, semantics and lexis.
6. Differentiate between functional and object-oriented programming (OOP).
7. Identify machine-code concepts and language.
8. Describe CISC and RISC instruction set architecture (ISA) families and their differences.
9. Define ASCII and Unicode, and interpreters and compilers.
10. Summarize scripting (e.g., Python, JavaScript, PowerShell and Bash).
11. Recognize virtual machines (e.g., NET, JAVA).
12. Describe the portability of code.
13. Identify high-level languages (e.g., C, C++, Java, Python, HTML and .NET), procedural and nonprocedural languages, and graphical language.
14. Describe an integrated development environment (IDE).

1.2 Software Basics

1.2.1 What is Software?

Technology, specifically, software, has become such an integral part of personal and work life that it may be hard to remember a world without it. Software makes difficult and complex tasks easy and simple, and impossible things possible.

The term software is defined as a collection of programs, procedures and rules that direct computer operations via lines of instructions. Software documentation instructs new users on how to operate an application, program or system properly.

A program, as defined by the International Organization for Standardization (ISO), is a sequence of instructions suitable for processing.[2] An assembler, compiler, interpreter or another translator is used to prepare the program for execution. The instructions may include required declarations and statements.

The American National Standards Institute (ANSI) defines a program as a set of one or more interrelated modules capable of being executed.[3] The terms routine, function and algorithm may be used when referring to parts of a program.

[2] US Food and Drug Administration, "Glossary of Computer System Software Development Terminology", 6 November 2014, https://www.fda.gov/inspections-compliance-enforcement-and-criminal-investigations/inspection-guides/glossary-computer-system-software-development-terminology-895

[3] *Ibid.*

Software development is the process by which user needs are translated into a usable software product—also called a software application because the software performs a specific task. The process involves:

1. Translating user needs into software requirements
2. Translating software requirements into software design
3. Implementing software design in program source code
4. Testing the source code
5. Installing the software application
6. Checking the software application for operational and business workflow activities
7. Maintaining the released software with bug fixes and new features that are released in incremental software upgrades

Figure 1.1 depicts the software development life cycle (SDLC).

Figure 1.1—Software Development Life Cycle

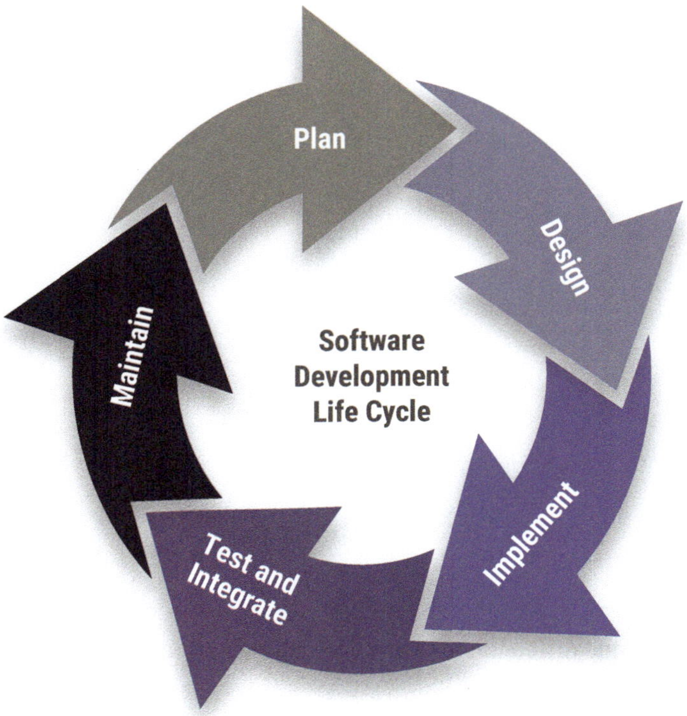

1.2.2 Software Types

Early software implementations were very dependent on hardware. However, with the advent of improvements in computer programming languages and hardware, today's software can be independent of hardware. There are four basic types of software:

- Firmware is also referred to as embedded software. Embedded system software controls a computer between the time it is turned on and the time the primary operating system takes control of the machine.[4] Firmware is also used to control hardware and devices that are not typically considered computers—telecommunications networks, cars, industrial robots and more.

[4] IEEE Standards Association, *IEEE 1275.2-1994 – IEEE Standard for Boot (Initialization Configuration) Firmware: Bus Supplement for IEEE 1496 (SBus)*, https://standards.ieee.org/standard/1275_2-1994.html

- System software provides the operating system and core functions, like disk management, utilities, hardware management and other operational necessities. Operating system examples include Windows and Unix for computers and iOS (Apple mobile operating system) and Android for mobile phones.
- Utility software supports the computer infrastructure. Examples include antivirus, file transfer and file management software, and programmer tools, such as text editors, compilers, linkers, and debuggers.
- Application software (application or app) helps users perform specific tasks. Office productivity suites, data management software, media players and security programs are examples.[5] Other application examples are web browsers, socializing platforms and music streaming software.

Figure 1.2 shows the layered relationship of the software types.

Figure 1.2—Software Types and Relationships

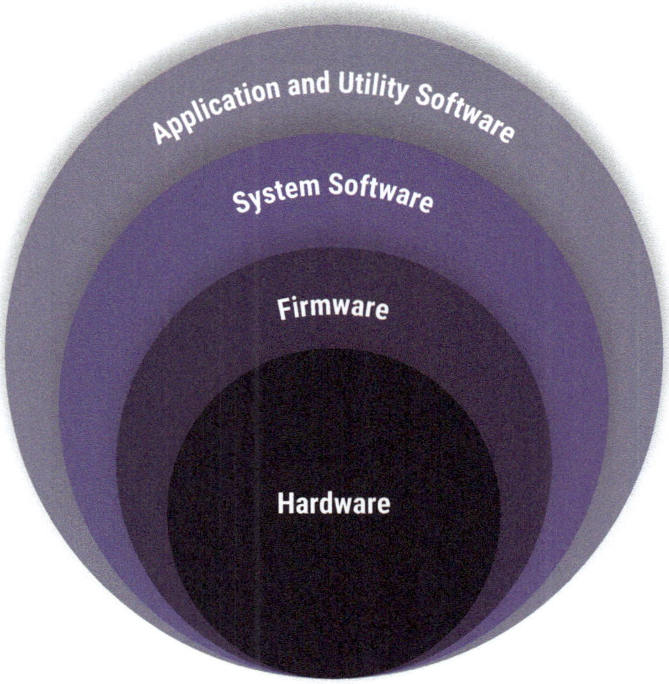

Figure 1.3 shows how to identify the different software types, and their common use and purpose.

Figure 1.3—Software Types, Use and Purpose

Software Type	Purpose	Examples
Firmware/embedded	Loading (or booting) the operating system, identifying the hardware configuration, testing and initializing the hardware, and providing interactive debugging facilities in case of faulty hardware or software.	BIOS (basic input output system) drivers, such as: • Display • USB • Motherboard • Sound card

[5] *Op cit* IBM, "What is Software Development?"

Figure 1.3—Software Types, Use and Purpose *(cont.)*

Software Type	Purpose	Examples
System software	Operating system, which is used to manage and optimize the behavior of the computer hardware with the software applications installed on it. System software is principally an intermediator, or a middle layer, between the user and the hardware.	• MS Windows • macOS • Linux/Unix • iOS • Android • Ubuntu
Utility software	Helps users configure, analyze, optimize or maintain a computer.	• Disk utilities, like defragmenters, compressors and cleaners • Antivirus programs • Registry cleaners and system restoration programs
Application software	Helps users perform work or leisure related activities.	• Office suite of products • Accounting application • Web browser • An email client

When writing a program, it is common to have errors in the program code. Debugging is determining the exact nature and location of a program error, and fixing the error. In software development, programmers have access to tools that allow them to set a breakpoint in the program to debug errors. A breakpoint is an intentional stopping or pausing in a program for debugging purposes. A stack trace is an audit trail of the method calls (i.e., instructions) that the application was compiling when an error occurred and can be used to debug the exact program line with the error.

When a computer system is initialized, or powered on, it begins by clearing memory and loading the operating system. In firmware, a boot program normally performs those functions and loads basic instructions that instruct the computer how to load programs into memory and how to begin executing those programs.

Two distinctions can be made between a warm boot and a cold boot:

- **Cold boot**—starting the system from a powered-down state. Included is a power-up self test, in which various portions of the hardware (such as memory) are tested for proper operation.
- **Warm boot**—restarting the computer while it is powered up. However, warm boot does not perform self tests, and it does not clear all memory.

1.2.3 History of Software

From basic tools to complex machinery, innovation is as old as humanity. Technology has long driven productivity improvements across all civilization, but the advent of computers and software increased the pace and scope of these improvements, at work and at home.

Figure 1.4 shows the major milestones of software history, categorized into logical eras.

Figure 1.4—Software History Milestones

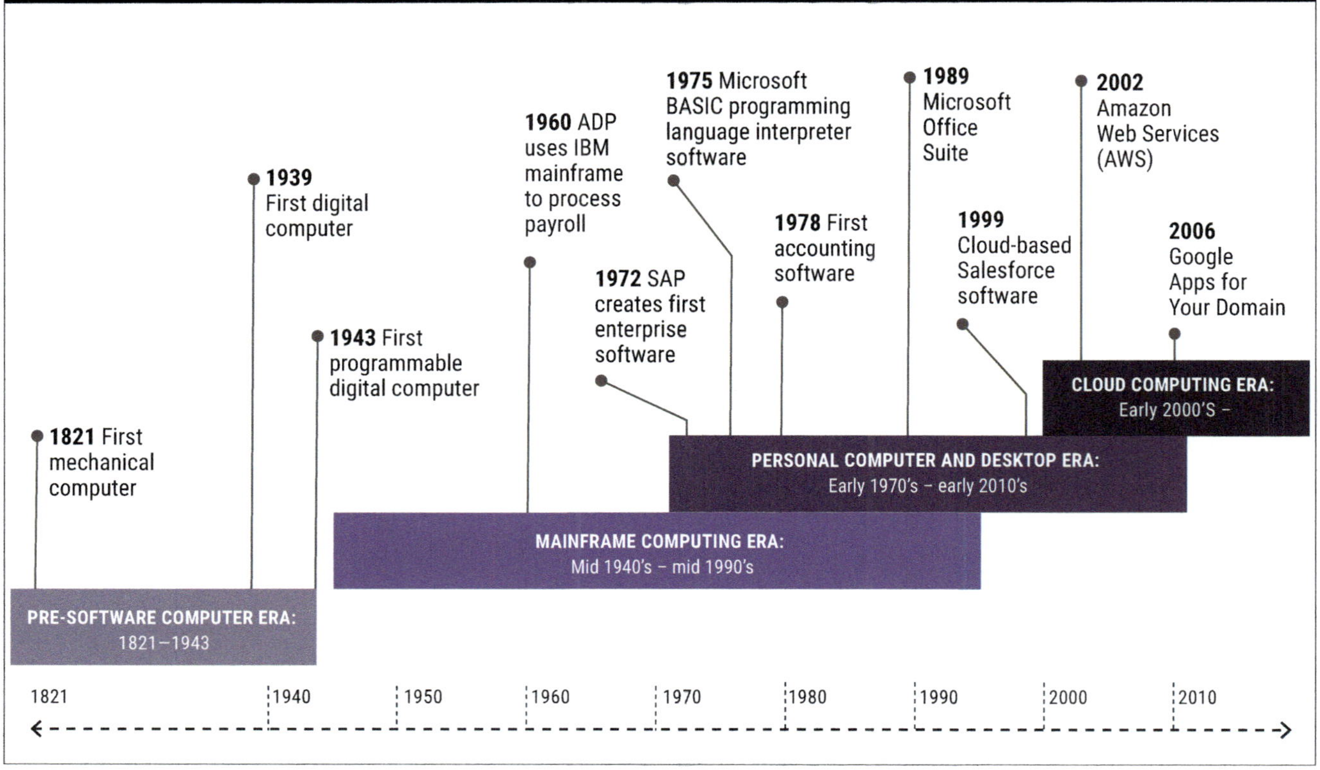

The pre-software computer era began with computers that were used mostly for large polynomial calculations. These computers completed in seconds the calculations that took hours for a human to perform.

The first mechanical computer was created in 1821 by Charles Babbage, an English mathematician who is known as the father of computing. He developed the difference engine, which used a series of gears and number wheels to compute polynomial functions.[6] The device consisted of 4000 parts and weighed over three metric tons.

A polynomial is a mathematical expression of two or more algebraic terms, that can contain constants, variables, coefficients and non-negative exponents. A common polynomial is the sum of several terms that contain different powers of the same variable. A variable can be a name, label, quantity or any defined data item whose value may be changed many times during processing. Typically, a single letter is used to represent a variable, e.g., x. Declaring a variable in a program allocates a memory location whose content may change during the program execution. In contrast, a constant is a value that does not change during processing and is sometimes referred to as hard-coded. Hard-coded is a term used when data or parameters in a program cannot be altered without modifying the program code.

IBM is the oldest software development and hardware manufacturer (over 100 years old) and was first in the field of information processing. All IBM products were developed and designed to record, process, communicate, store and retrieve information—from its first scales, tabulators and clocks to today's powerful computers and vast global networks.[7]

The standard punched card was invented by Herman Hollerith in the 1800s, before IBM was incorporated. In the late 1800s, the data stored on punched cards were tabulated for use by the census in New York City. The punched card idea originated in the textile industry; Jacquard punched cards came from the idea to use a connected chain of punched wooden cards to guide rug-making looms.

6 University of Minnesota, "Who Was Charles Babbage?," Charles Babbage Institute, 9 February 2016, www.cbi.umn.edu/about/babbage.html

7 IBM, "Chronological History of IBM," IBM Archives, www.ibm.com/ibm/history/history/history_intro.html

A modern example of a punch card is the common scantron, which is used in the educational sector for testing. **Figure 1.5** shows that each card circle is a bit of data that translates to 1, true or on in computer language when the circle is shaded by a pencil, or translates to 0, false or off when the circle is not shaded.

Figure 1.5 shows a sample scantron.

Figure 1.5—Sample Scantron

Source: ImagePixel; Getty Images

The 1940s is the beginning of digital electromagnetic, programmable computers, the foundation of software today. During World War II, the Germans used a machine named Enigma to send and receive coded messages to and from the field. Enigma and the more sophisticated Lorenz machines used an electromechanical rotor to scramble the 26 letters of the alphabet, thus obscuring the message or ciphering it.

Also during this era:

- Alan Turing, known as the father of computer science, invented the Bombe machine to break the code of Enigma. This complex machine consisted of approximately 100 rotating drums, 10 miles of wire and about one-million soldered connections.
- Colossus computers at Bletchley Park, in the United Kingdom, were the first digital electronic computers. They helped decode German messages during World War II. Colossus computers used thermionic valves to perform Boolean and counting operations. Boolean pertains to the principles of mathematical logic developed by George Boole, a nineteenth century mathematician. Boolean algebra is a type of mathematical operation carried out on variables that can have only one of two possible values, i.e., 1 (true) and 0 (false). Like add, subtract, multiply and divide are the primary operations of arithmetic, and, or and not are the primary operations of Boolean logic.

  Colossus is regarded as the first programmable, electronic, digital computer, although it was programmed by switches and plugs and not by a stored program. The Colossus programmers were 273 members of the Women's Royal Naval Service, referred to as Wrens, who received training in binary math, the teleprinter alphabet and how to read machine punch tapes.

The cryptographers, led by Alan Turing, decided which patterns to run, and the Wrens entered the desired programming into Colossus using switches and telephone-exchange plugs and cords. This process was called pegging a wheel pattern.[8]

- The Electronic Numerical Integrator and Computer (ENIAC), built from 1943 to 1946, is regarded as the first general-purpose electronic digital computer. The military used ENIAC to calculate ballistic tables.[9] ENIAC contained 20 electronic accumulators, each of which could store a 10-digit decimal number (**figure 1.6**).

  Its logic circuits were also electronic. The sequence of operations was set by the placement of patchcords in plugboards. ENIAC had read-only memory of about 300 numbers, which were entered by turning switches. An IBM card reader and an IBM card punch provided input and output. ENIAC contained 18,000 tubes, 70,000 resistors, 10,000 capacitors, 1500 relays and 6000 manual switches.[10]

  This equipment consumed 140 kW of power and filled a 20-foot by 40-foot room.

Figure 1.6—ENIAC Decade Ring Counter

Replacing a bad tube meant checking among ENIAC's 19,000 possibilities.

Source: Wikipedia, https://en.m.wikipedia.org/wiki/File:ENIAC-changing_a_tube.jpg

The operations of the Bombe, Colossus and ENIAC computers were hard-wired, moving levers and parts, and required individuals to operate a sequence of very specific tasks to obtain results. In later software eras, these steps were converted to stored programs that were known as soft-ware, compared to the hard, mechanical counterparts that were known as hard-ware.

The now-outmoded terms of cipher and code breaking algorithm have been replaced with the terms encryption and decryption, respectively. The Bombe, Colossus and ENIAC were able to decrypt Enigma code, due to its weak encryption elements: the key permutations were limited to a predetermined length of 26 letters, 3 wheels and 13 cables. The longer the key, the more difficult it is to compromise in a brute force attack where all possible key combinations are tried.[11]

[8] Marsh, A.; "The Hidden Figures Behind Bletchley Park's Code-Breaking Colossus," *IEEE Spectrum*, 31 December 2019, https://spectrum.ieee.org/tech-history/dawn-of-electronics/the-hidden-figures-behind-bletchley-parks-codebreaking-colossus

[9] Engineering and Technology History Wiki, "ENIAC," 23 November 2017, https://ethw.org/ENIAC

[10] *Ibid.*

[11] ISACA, *Cybersecurity Fundamentals Study Guide, 3rd Edition*, USA, 2021

1.2.4 Mainframe Era

The mainframe era, during and post World War II, was geared towards enterprises because the computers were very large and astronomically costly to build, operate and maintain. The knowledge to operate these computers was uncommon and not many people were skilled in mathematics and binary calculations.

The meaning of mainframe changed over the years. Mainframe is the concatenation of two words that described a part of the system for early computers—the main frame was the term for the metal housing for the electronic components that performed the computing-instruction processing, and arithmetic and held memory. Other frames contained disk, tape, communications, etc.

Mainframes continue to power backend systems that run the economy today—in particular, at banks, insurance companies, airlines and a myriad of federal and state government agencies.[12] Until the mid-1990s, mainframes provided the only acceptable means of handling the data processing requirements of a large business. These requirements are based on running large and complex programs, such as payroll and general ledger processing.

The mainframe owes much of its popularity and longevity to its inherent reliability and stability, a result of careful and steady technological advances made since the introduction of the IBM® System/360 in 1964. No other computer architecture can claim as much continuous, evolutionary improvement, while maintaining compatibility with previous releases.

Because of these design strengths, the mainframe is often used by IT organizations to host the most important, mission-critical applications. These applications typically include customer order processing, financial transactions, production and inventory control, payroll and many other types of work.

Enterprises use mainframes for applications that depend on scalability and reliability. For example, a banking institution can use a mainframe to host the database of its customer accounts, for which transactions can be submitted from any of thousands of ATM locations worldwide.

Businesses today rely on the mainframe to:

- Perform large-scale transaction processing (thousands of transactions per second)
- Support thousands of users and application programs concurrently accessing numerous resources
- Manage terabytes of information in databases
- Handle large-bandwidth communication[13]

1.2.5 Personal Computer and Desktop Software Era

The idea of the personal computer began to infiltrate the market and people's homes in the 1970s, when Intel introduced the small microprocessor and hardware costs dropped. A microprocessor is a type of integrated circuit or chip and is at the heart of every computer today, and, thus, began the rise of the personal computer and desktop software era.

The microprocessor shrunk the size and cost of a computer so that it could fit on a desktop and was affordable, which allowed computer enthusiasts to purchase their own personal computer. A milestone was the Altair microcomputer kit, first offered for sale in 1975. One of the Altair designers, Edward Roberts, hired two young programmers named William Gates and Paul Allen to write software for it. Gates and Allen later formed Micro-Soft (now Microsoft), a

[12] Rafter, M.; "Mainframes Are Having a Moment," *IEEE Spectrum*, 17 April 2020, https://spectrum.ieee.org/tech-talk/computing/software/mainframes-programming-language-cobol-news-coronavirus

[13] IBM, "Who uses mainframes and why do they do it?," IBM Knowledge Center, July 2020, www.ibm.com/support/knowledgecenter/zosbasics/com.ibm.zos.zmainframe/zconc_whousesmf.htm

company that led the development of personal computer software[14] and opened possibilities to individuals and smaller businesses.

In 1976, the Apple computer company offered a competing kit-form product, with a different architecture. In 1977, the company introduced the Apple II, which had a sleek beige plastic case and a color monitor, and used a microprocessor made by Motorola.[15]

1.2.6 Cloud Computing Era

Software as a Service (SaaS) is software that resides (hosted) in a cloud architecture. Cloud computing provides convenient, on-demand network access to a shared pool of resources that can be rapidly provisioned and released with minimal management effort or service provider interaction. In a cloud computing environment, software users usually access data using a web browser and a thin client. A thin client is a lightweight computer that is optimized for establishing a remote connection with a server-based computing environment by using a browser instead of a local software application.

The server side does all the work, which can include launching software programs, performing calculations and storing data. A server is a high-speed computer in a network that is shared by multiple users and holds the programs and data that are shared by all users, e.g., a server runs a single instance of software that services numerous users.[16] A server is also defined as a dedicated computer or device that manages specific resources (e.g., a file server, web server or an enterprise server).

[14] Engineering and Technology Wiki, "The First PCs," 24 October 2019, https://ethw.org/The_First_PCs

[15] *Ibid.*

[16] Ali, A.Q; A.B.M. Sultan; A.A.A. Ghani; H. Zulzalil; "A Systematic Mapping Study on the Customization Solutions of Software as a Service Applications," *IEEE Access*, vol. 7, pp. 88196-88217, 2019

Figure 1.7 depicts the architecture of a server-client relationship.

Figure 1.7—Architecture of Server-Client Relationship

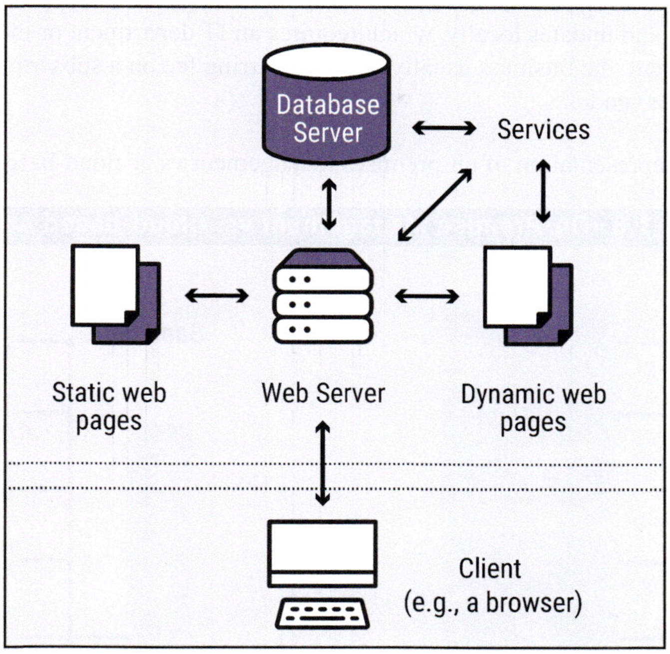

The high regard for SaaS in the business domain can be attributed to several factors:

- **Cost-efficiency**—Building their own servers and tools is time-consuming and expensive for businesses, because they need to order, pay for, install and configure expensive hardware. However, by using cloud computing, an enterprise only pays for the amount cloud computing processing power it uses. Moreover, with cloud computing, costly software licenses and long-term upgrade fees are replaced by lower monthly subscription costs, which include all future software updates. Cloud computing offers cost savings.
- **Reliability**—Also known as uptime, reliability provides much more available, dependable and consistent service than any in-house IT infrastructure. Reliability guarantees 24/7 and 365 days of service. If any of the servers fail, then hosted applications and services can easily be transitioned to other available servers.
- **Unlimited Storage**—Cloud computing provides almost unlimited storage capacity; the provider has a multitude of servers that offer storage space à la carte. An enterprise does not need to purchase upfront or anticipate the storage for the future, SaaS allows for complete customization and cost savings growing *with* the business needs versus predefined, rigid plans. Storage can be accessed, added to and removed on an as-needed basis.
- **Backup and Recovery**—When storing data in the cloud, backing it up and restoring it is relatively easier than storing it on a physical device, because the hosting cloud architecture usually provides all needed software to accomplish those tasks and, in most cases, automate them.
- **Easy Access to Information**—When users register in cloud, they can access their accounts from anywhere in the world, provided there is Internet connection at that location. Storage and security considerations vary depending on the account type chosen and the fees one is willing to pay.
- **Security**—Although security can be much more robust compared to small business capability, security is highly dependent on the deployment model chosen and the specific configurations. Concurrently, with the expansion of cloud usage, cloud providers are facing an increased risk of attacks and breaches, due to the accumulation and clustering of sensitive data and information in a centralized server environment.

  Enterprises experiencing a loss of sensitive data from the cloud may not be in a position to bring an action against the perpetrators, because the cloud provider often has to initiate legal action.[17]

[17] *Op cit* ISACA, *Cybersecurity Fundamental Study Guide 3rd Edition*

Multi-tenancy is a vital feature of cloud computing.[18] Multi-tenancy allows a single instance of software running on a server to service numerous businesses (tenants).

In the on-premises model of software architecture the business manages all hardware, software, network, data and data storage, maintenance, and updates locally, which requires an IT department or individual resources. In a managed, hosted environment, the business usually pays a recurring fee on a subscription basis to transfer those responsibilities and risk to a vendor.

Figure 1.8 is a graphical representation of on-premises management vs. a cloud-based architecture.

Figure 1.8—On-Premise Management vs. Cloud-based Architecture

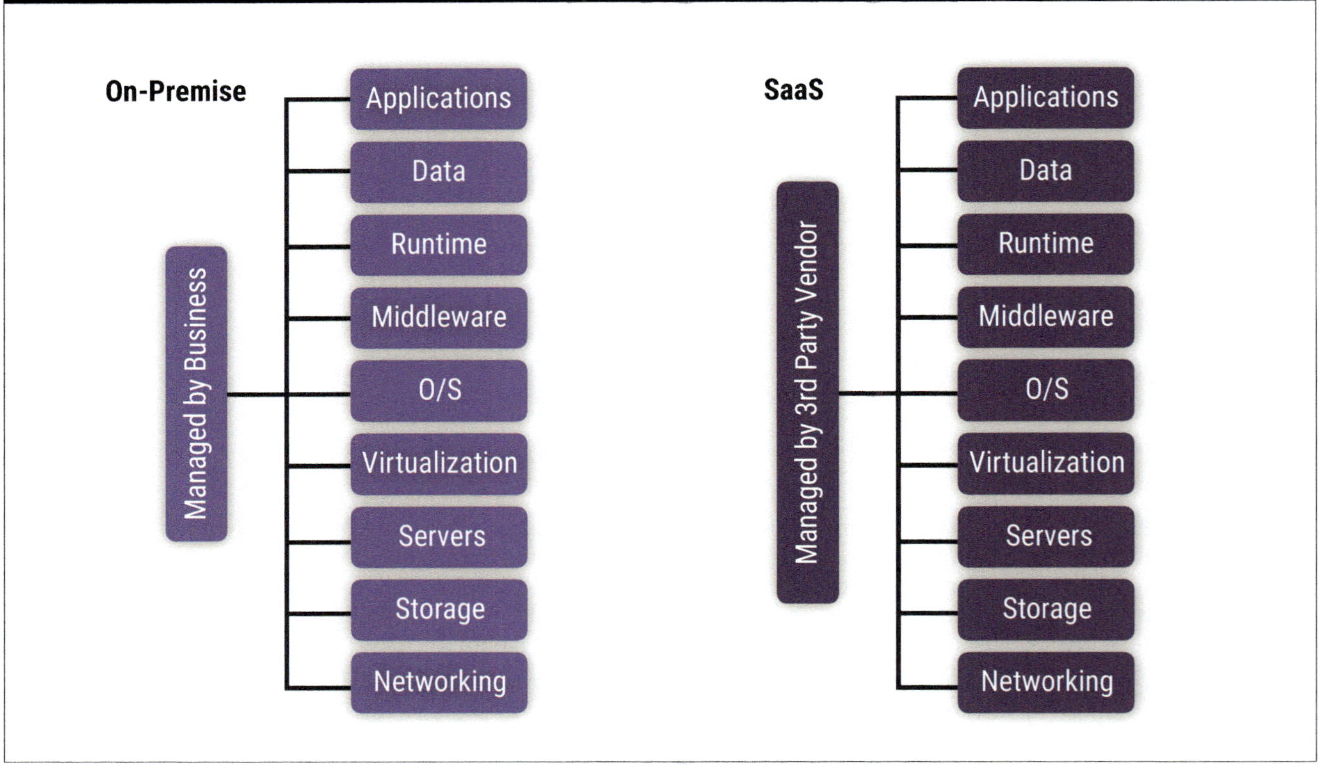

1.2.7 Y2K Bug

The notorious Y2K bug (in metric measurements, K represents one thousand), also called Millennium Bug or Year 2000 bug, is a well-known problem in the coding of computerized systems that was projected to create havoc in computers and computer networks when the world transitioned to a new millennium, from December 31, 1999, to January 1, 2000.

Before 2000, computers were coded to abbreviate the four-digit year, e.g., 1999, as two digits, e.g., 99, to save on memory allocation space, because computer hardware, and especially memory, was astronomically costly. If the computer year allocation was two digits, then when the clock turned to 2000, it was feared that the last two digits, i.e., 00, would be understood by the computers as 1900, and not 2000.

[18] *Op cit* Ali

Enterprises and governments around the world anxiously prepared for the year 2000, because they feared that all computer-dependent systems, including financial, medical and military systems, would cease to function or, at a minimum, function incorrectly.

Avoiding the Y2K bug required one of two possible options: entirely rewrite the whole program to use a four-digit year, or adopt a quick, temporary patch fix called windowing. Most installed the temporary patch. Windowing is a temporary patch that treats all dates ending with double zeroes as the start year 2000, rather than 1900. This twist of logic to trick computers only works for a few decades—typically 30 years.[19]

When windowing logic is applied, the programmer instructs the software to use the twenty-first century for dates that fall within a specific date range—a window of time. The computer processes the year based on a future date that the programmers have set as the assumption decade (e.g., twenty-first century instead of twentieth century), known as program pivot. For example, setting a pivot of 30 in the year 2000 is sustainable from 2000 through to 2029; then, in 2030, the program will revert to assuming that the century is 1930, and not 2030. The arbitrarily set pivots can cause other backward compatibility issues, such as when setting the pivot to 70 years in the year 2000 may cause issues with birth dates earlier than 1969.

Because programmers addressed the potential Y2K issues, very few problems occurred in computer systems when the clock reached the year 2000. Enterprises that chose the temporary patch fix found it to be acceptable, because it was sustainable for three decades and legacy computer systems are eventually replaced by new technology.[20]

1.2.8 Algorithms

An algorithm is a finite set of well-defined, unambiguous rules for the solution of a problem in a finite number of steps. It is a sequence of operational actions that leads to a desired goal and is the basic building block of a program.

The further that the language of a program is away from machine language, the slower the program executes and the more memory it requires. To be able to use natural language and write logic that is easier, faster and portable, programmers must write clear and efficient algorithms (which make up the logic) that are not bloated with steps that can be avoided.

1.2.9 Differences Among Syntax, Semantics and Lexis

In the same way that spoken languages have structure, spelling and meaningful words to communicate properly, a computer language has its own vocabulary, spelling, form and structure.

Programmers (i.e., developers, engineers and coders) write software applications using a set of instructions comprised of arguments (i.e., parameters and variables) and a set of available control flows that the program uses to manipulate the data via logical statements and program semantics to output a desired result.

The three main components of a programming language are described as follows:

- Syntax in a program is the form or structure of the expressions, statements and program units. Syntax can be likened to grammar of a sentence. In code, finding the error of a missing bracket ({}), statement terminator (;) or a comma (,) can be very time consuming. However, today's compilers and integrated development environments (IDEs) are much more advanced and highlight the line missing the correct syntax and fail the compilation.

[19] Buchner, M.; "Windowing as a Y2K Solution," *Enterprise Systems Journal*, 5 May 1999, https://esj.com/articles/1999/05/24/ad-trends—windowing-as-a-y2k-solution.aspx
[20] *Ibid.*

- Semantics is the computational thinking analysis of the written code. If a programmatic statement is written semantically wrong the logical operation may be compiled (or not), but the result will not have the intended meaning.

Example: The expressions 2+3*5 and (2+3)*5 have different meanings.

- Lexis is the entire set of a language's vocabulary. A misuse or misspelling of identifiers, operators and keywords that may apply to another language but are not recognized in the program in which they are used is a lexis error.

1.2.10 Developing an Algorithm

An algorithm is a finite set of well-defined phases toward the solution of a problem in a finite number of steps. The typical process of creating an algorithm is a collection of six phases beginning with identifying the problem and ending with documenting the solution.

Example: A property management company is experiencing challenges with keeping cash in the office and wants to receive payments electronically. Considering the following algorithmic phases, what are the specific steps, at a high level, that are needed to outline this algorithm?

Figure 1.9 shows the typical phases in the development of algorithms.

Figure 1.9—Steps in Algorithm Development

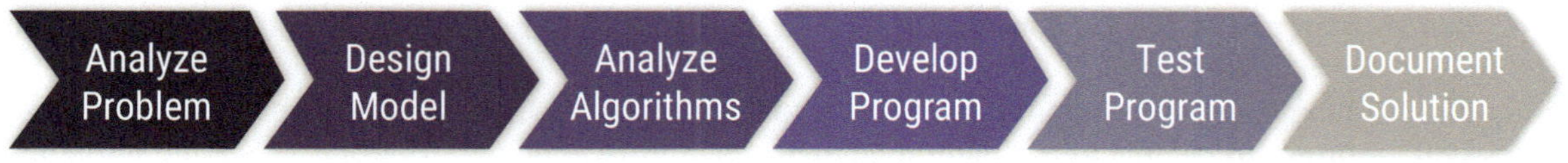

1.2.11 Analyze Problem

One of the most important phases is defining and analyzing the problem. Asking lots of questions about the current problematic or undesired process can uncover solutions or bring attention to unknown things not included in the initial problem statement.

To help in this initial analysis, this step is divided into three separate components:

- **Input**—A list of the source data provided to the problem, e.g., three numbers that need to be added
- **Processing**—A list of actions needed to produce the required outputs, e.g., read the numbers, add the numbers
- **Output**—A list of the output required (desired goal), e.g., sum of the three numbers, print results

1.2.12 Design Model

An algorithm can be communicated and documented through textual steps; a graphical representation, such as a box diagram; input-process-output chart; state diagram; unified modeling language (UML); control flow diagrams; pseudocode or flowcharts to convey the analysis steps.

A control flow diagram is a visual, graphical representational diagram that depicts the set of all possible sequences in which operations may be performed during the execution of a system or program.

Pseudocode syntax is also an option that is often used. Pseudocode is a combination of programming language constructs and natural language used to express a computer program design (**figure 1.10**). Generally, it is used to introduce computer science students to algorithms in a programming course and to reinforce the idea that program design should precede code.[21] Pseudocode cannot be executed on a computing platform.[22]

Figure 1.10—Pseudocode

Pseudocode

1. Start
2. Ask for input of a, b, c
3. Read a, b, c
4. d=a+b+c
5. Print or display d
6. Stop

Flowchart

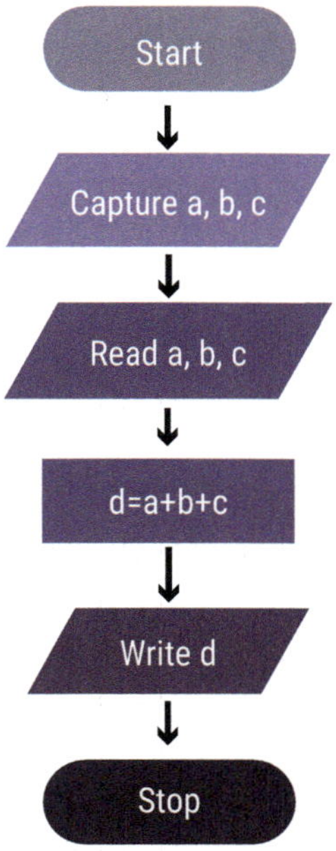

Program code (e.g. Java)

```
//Program to find the sum of three numbers

Import java.util.Scanner;
Public class, Add{
        public static void main(String [] args){
        scanner Ob1 = new Scanner(System.in);

System.out.println("Enter value of a:");
Int a = Ob1.nextInt();

System.out.println("Enter value of b:");
Int b = Ob1.nextInt();

System.out.println("Enter value of c:");
Int c = Ob1.nextInt();
Int d = a+b+c;

System.out.println("Sum of given three
numbers is: "+d;
        }
}
```

Unified modeling language and flowcharts differ in one way; a flowchart is an informal term used to define a control flow diagram in which suitably annotated geometrical figures are used to represent operations, data, or equipment and arrows are used to indicate the sequential flow from one to another. Informally, it can use any shape to show the sequence of the steps. UML, however, is an ISO/IEC (19501:2005) standard modeling language where each notation and shape has its own inherent meaning (**figure 1.11**).

Figure 1.11—Shapes for UML Flowcharts

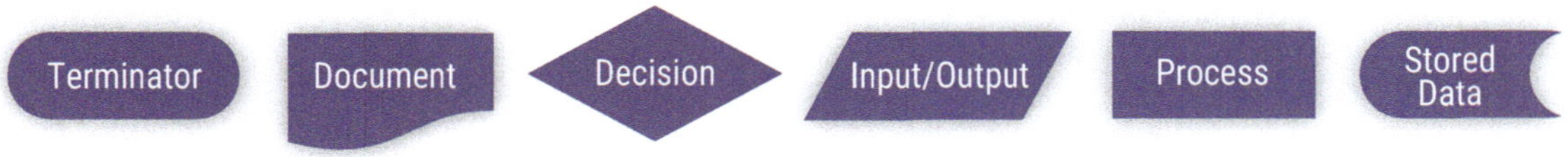

[21] Olsen, A.L; "Using pseudocode to teach problem solving," *J. Computing Sciences in Colleges*, 2005, p. 235

[22] Dirgahayu, T.; S.N. Huda; Z. Zukhri; C.I. Ratnasari; "Automatic translation from pseudocode to source code: A conceptual-metamodel approach," 2017 IEEE International Conference on Cybernetics and Computational Intelligence (CyberneticsCom), Phuket, 2017, p. 122.

1.2.13 Analyze Algorithm

Some algorithms perform precise mathematical functions, such as sorting, finding the maximum of a sequence of numbers, determining primality or finding the square root. Much is known about optimal mathematical logic behavior, particularly with respect to the two important parameters of execution time and data storage requirements (e.g., memory).

Algorithm analysis is used to predict performance (e.g., time to launch an application), compare algorithms (e.g., best method to solve a problem and industry standards) and provide a level of benchmarking for the next steps in the software development life cycle (e.g., testing and integration).

There are several algorithmic analysis exercises to employ, depending on the problem resolution: theory algorithm, observational methods, mathematical models, order of growth classification and memory.

Theory algorithm is used to identify the best- and worst-case scenarios and to help the developer choose the best method to solve the problem and predict performance. This algorithmic analysis exercise considers three types of analyses:

- Best-case scenario (also known as happy path)
- Worst-case scenario
- Average case

One of the most important aspects of algorithm design is to create an algorithm that has an efficient runtime. **Figure 1.12** displays the commonly used notations that assist the developer in identifying the algorithm:

- **Upper bound**—Performance maximum benchmark
- **Lower bound**—Performance minimum benchmark
- **Ideal behavior to within a constant factor**—A constant factor does not depend on the input parameter(s) and remains constant (e.g., the amount of time an algorithm takes to execute is dependent on processor speed, disk writing speed, compiler, etc.)

Figure 1.12—Commonly Used Notation to Identify Algorithm Bounds and Behavior

Notation	Provides:	Example	Shorthand for:	Used to:
Big Theta	Asymptotic order of growth	$\Theta(N^2)$	$\frac{1}{2} N^2$ $10 N^2$ $5^2 + 22 N \log N + 3N$ ⋮	Classify algorithms
Big Oh	$\Theta(N^2)$ and smaller	$O(N^2)$	$10 N^2$ $100 N$ $22 N \log N + 3 N$ ⋮	Develop upper bounds
Big Omega	$\Theta(N^2)$ and larger	$\Omega(N^2)$	$\frac{1}{2} N^2$ N^5 $N^3 + 22 N \log N + 3 N$ ⋮	Develop lower bounds

Source: Sedgewick, R.; K. Wayne; *Algorithms, 4th Edition*, USA, 2020, https://algs4.cs.princeton.edu/lectures/keynote/14AnalysisOfAlgorithms-2x2.pdf

The notation is often used to express a relative representation of the complexity of an algorithm with the data input.

The process of algorithm analysis ensures high quality of the written function by comparing similar functions. Analysis further confirms that algorithms selected are correct, appropriate and stable, and meet all accuracy, timing and sizing requirements.

Although some programs challenge defined specifications, it is the responsibility of the programmer to estimate, and always attempt to improve, their algorithmic logic and processing performance. In the property management company example, the test would be to add very large numbers for each of the requested variables. The running time can vary widely depending on the input[23] and the complexity of the calculation.

Observational algorithm analysis can be used to measure runtime of an algorithm via empirical analysis, either manually or automatically, by using various inputs and measuring the time to process (**figure 1.13**).

Figure 1.13—Sample Measurement of Process Times

N	Time (seconds)
250	0.0
500	0.0
1,000	0.1
2,000	0.8
4,000	6.4
8,000	51.1
16,000	?

Source: Sedgewick, R.; K. Wayne.; *Algorithms, 4th Edition*, USA, 2020, https://algs4.cs.princeton.edu/lectures/keynote/14AnalysisOfAlgorithms-2x2.pdf

Data analysis of the input vs. runtime can be accomplished via a standard plotting chart or log-log plot (**figure 1.14**) to graphically display the exponential growth of the relationship between input and time.

Figure 1.14—Example Input vs. Runtime

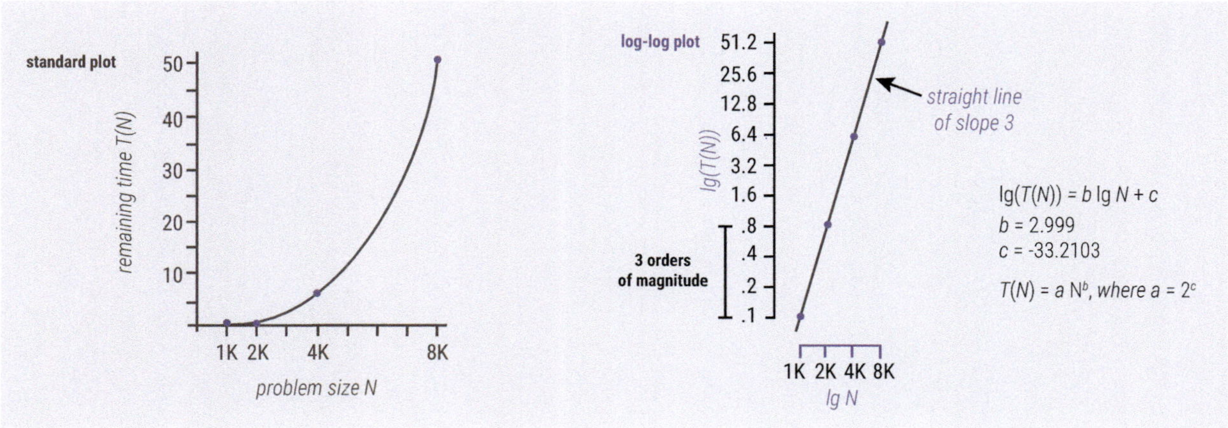

Source: Sedgewick, R.; K. Wayne; *Algorithms, 4th Edition*, USA, 2020, https://algs4.cs.princeton.edu/lectures/keynote/14AnalysisOfAlgorithms-2x2.pdf

[23] Sedgewick, R.; K. Wayne; *Algorithms, 4th Edition* (online), "1.4 Analysis of Algorithms," 17 July 2020, https://algs4.cs.princeton.edu/14analysis/

To capture and predict performance of an algorithm, a doubling hypothesis can be used to measure the input and runtime relationship by doubling the size of the input for each test run to identify results. This allows for quicker identification of any bounds, out of bounds, or constant association in shorter time and a lesser number of tests (i.e., executions).

Each instance of a calculation, regardless of how simplistic, within an algorithm costs time and memory to process and greatly affects the user experience, type of machine and its processing power, and any timeout behaviors that may be experienced. **Figure 1.15** depicts an example of simple operations and execution times on a MacBook Pro 2.2GHz with only 2GB of RAM, on an OS X operating system.

Figure 1.15—Example Operations and Execution Times

Operation	Example	Nanoseconds
Integer add	a + b	2.1
Integer multiply	a * b	2.4
Integer divide	a / b	5.4
Floating-point add	a + b	4.6
Floating-point multiply	a * b	4.2
Floating-point divide	a / b	13.5
Sine	Math.sin(theta)	91.3
Arctangent	Math.atan(y, x)	129.0
...	...	...

Source: Sedgewick, R.; K. Wayne; *Algorithms, 4th Edition*, USA, 2020, https://algs4.cs.princeton.edu/lectures/keynote/14AnalysisOfAlgorithms-2x2.pdf

All variables must be considered when writing an algorithm, to consider the experience of most users that will use the application. That consideration includes variables outside of the developer's control, ensuring, at a minimum, no catastrophic program behavior, such as a program crash or loss of data.

Figure 1.16 shows an example of doubling hypothesis testing.

Figure 1.16—Double Hypothesis Testing

N	time (seconds) †	ratio	lg ratio
250	0.0		–
500	0.0	4.8	2.3
1,000	0.1	6.9	2.8
2,000	0.8	7.7	2.9
4,000	6.4	8.0	3.0
8,000	51.1	8.0	3.0

$$\frac{T(2N)}{T(N)} = \frac{a(2N)^b}{aN^b} = 2^b$$

lg (6.4 / 0.8) = 3.0

seems to converge to a constant b ≈ 3

Source: Sedgewick, R.; K. Wayne; *Algorithms, 4th Edition*, USA, 2020, https://algs4.cs.princeton.edu/lectures/keynote/14AnalysisOfAlgorithms-2x2.pdf

Moreover, algorithmic analysis is not only for mathematical computations. When a programmatic decision is made about which conditional statements to use (i.e., control flow structures, such as if and while that are used in programming code to make decisions—explained in section 2.3.3 and following), it is best to always try to write the least amount of code in the clearest, shortest way.

Spaghetti code is a colloquial term used often to refer to badly written code that is convoluted and difficult to read and maintain. It is extremely rare to have code that requires no updates, so careful analysis at this stage is a valuable exercise.

1.2.14 Develop Program

After the solution outline is accepted and analysis is completed, the pseudocode, charts or text are translated to source code using the vocabulary and syntax of a programming language.

1.2.15 Test Program

At this phase, the program is tested by the programmer via low-level tests, software testers, the end user or all three. Testing can cover all areas of the program: user interface, data entry and output, integrations, security, performance, and much more.

One test case example is entering different numbers to attempt to break the code, e.g., if the input data type is an integer, then entering a decimal causes the program to either stop and exit, get hung up or result in an unhandled exception. An unhandled exception occurs when a program attempts to use (via input) or access data (via memory) incorrectly. In simplest terms, it means that the developer did not write code to handle certain use cases and scenarios for each program branch of decision making (covered in greater detail in Chapter 2: Programming).

1.2.16 Document Solution

Documentation has two parts: code documentation and user documentation.

Documenting the code logic is critical for programmers to be able to read and understand the program code if any part of the code needs modification in the future. For example, code documentation should detail function purpose, expected outcome and variables.

Explanatory text can be inserted into a program instruction set in comments, without having any effect on the execution of the program. Comment text provides clarification to human readers, but does not affect machine interpretation. Commented text is noted with delimiters that vary among programming languages.

Each programming language has its own notation for comments. **Figure 1.17** gives a few examples of how comments are indicated in programming languages.

Figure 1.17—Comment Syntax in Programming Languages	
Program Language	Comment Syntax
C and C++ comment variations	Single inline comment: // Example: accBal //example inline comment Multiline comment: /* */ Example: /*this is a Multiline comment*/ Comment block: /************************\ * * *. * ************************/ Example: /************************\ *This is an example comment block* *This is an example comment block* ************************/ Multiline comment option by Allen Holub: /* */ /* Example: /* This is a multiline style recommended * by Allen Holub * a software architect and author. */ There are many other comment styles from which to choose.
Python	Single inline comment: # Example: # This hashtag can be used to comment inline Python does not support multi line comments. A hashtag is needed on each comment line.

Figure 1.17—Comment Syntax in Programming Languages *(cont.)*	
Program Language	**Comment Syntax**
Perl	Single inline comment: # Example: # This hashtag can be used to comment inline Multiline comment: Multiline comments in Perl are created using Plain Old Documentation (POD). =pod This is a multiline comment =cut
Windows PowerShell	Single inline comment: # Example: # This hashtag can be used to comment inline Multiline comment: <# #> Example: <# multi-line comment #>
BASIC	Single-line comment: REM Example: REM this line is a remark (comment)
AppleScript	Single-line and multiline comment: (* *) Example: (* This is a comment in AppleScript *) .
AppleScript and Structured Query Language (SQL)	Single inline comment: — Example: —A comment in AppleScript and SQL

Figure 1.17—Comment Syntax in Programming Languages *(cont.)*	
Program Language	**Comment Syntax**
Java	Single inline comment: // Example: // This is an inline comment in Java Comment block: /** * * */ Example: /** * This is a block comment in Java * This is another line of block comment * The comment block can be as long and as wide as needed */

Furthermore, using practical, relevant naming conventions is good practice and is extremely helpful for the current developer and any future developers tasked with maintaining that program.

Today's programs aim to be self-explanatory by designing user interfaces (UI) that make interacting with the application easy for users, which results in an improved user experience (UX), thus reducing the need for long, robust user documentation guides.

Although user documentation is not as prevalent as it once was, it is still good practice to include documentation, or, at least a knowledge base and FAQ (Frequently Asked Questions) section online.

1.2.17 Functional vs. Object-Oriented Programming (OOP)

Functional programming is designed to handle symbolic computation and list processing applications and is based on mathematical functions. Functional programming is building programs (i.e., designing, architecting) based on fundamental software construction. It focuses on declarative (what to solve) programming rather than imperative (how to solve), and application state flows through pure functions, thus avoiding shared states, mutable data and side effects. Following are functional programming core concepts:

- **Pure function**—This is the simplest practical core concept. Pure function is a function that if the input is the same, it returns the same value; without modifying another variable, argument, or output (i.e., no side effects).
- **Function composition**—Composition of functions is the process of combining two functions to produce a third function with different behavior and a computation or output. In other words, the composition of function f with the function g is written f(g(x)), which represents the function f of the function g of x. This composition function applies function g to the independent variable x, and then applies the second function f to the result of function g. In mathematics and programming functions, of is sometimes represented as the letter o, to be read as f of s of x (e.g., (f o s)(x)). The function s is inside the function f.

The mathematical term PEMDAS is important to understand as the basic order of operations. If PEMDAS is not followed, the result can be incorrect computation and outputs. PEMDAS is an acronym for the order of operations: parenthesis, exponents, multiplication, division, addition, subtraction. PEMDAS can easily be remembered with this mnemonic: **P**lease **E**xcuse **M**y **D**ear **A**unt **S**ally. **Example:** $3^2 + (2+12x2) * 16/4$

The order of operations for the example is:

1. 12x2=24
2. 2+24=26
3. 3^2=9
4. 26x16=416
5. 416/4=104
6. 9+104=113

- **Shared state**—Shared state is when a variable, object, attribute or memory states are shared, much like inheritance in OOP. Functional programming avoids shared state because it can be very complex and error prone; the variables, objects and memory require traceability to identify which function, or order of operation, propagated the initial error that resulted in an erroneous final output.
- **Mutating state**—An object can either be a variable, often referred to as mutable or changeable after it is created, or immutable, which is a static object that cannot be changed. Immutability is a central idea in functional programming because it dictates the data flow.
- **Side effects**—A side effect is any programming application state change that can be seen from *outside* the called function except for the intended return value of the function. Side effects can be expressed via any of the following:
  - Output to the screen
  - Writing to a file
  - Triggering any external process
  - Calling any other functions
  - Logging

The main benefit of functional programming is the modular approach of functions (subprograms). Separating side effects from the rest of the logic (code) can make a program easier to maintain, test, debug, extend and refactor. Some popular functional programming languages still used today include Lisp (developed at Massachusetts Institute of Technology and used mostly for artificial intelligence) and Python.

Object-oriented programming (OOP) bundles everything as self-sustainable objects. Considering software development design architecture, OOP requires the developer to break down the problem into logical siloed **objects** that perform a complete task independent of the rest of the program. A design/architectural process good practice is to identify all the objects the programmer wants to manipulate and how they relate to each other, prior to coding; this is known as data modeling.

A class in OOP is a blueprint, or template, to create an object. A class has properties and methods. An object is an instance of a class. Object-oriented programming contains the following four pillars:

- **Inheritance**—A a mechanism where a class can be derived from another class for a hierarchy of classes that share a set of attributes and methods (i.e., parent-child relationship where attributes are inherited).
- **Polymorphism**—Describes the condition of an occurrence in different forms. The ability to use a single-type entity (method, operator or object) to represent different entity types in different scenarios, extending flexibility of existing code. Two additional representations regarding polymorphism are:
  - **Function overloading**—Allows the programmer to write functions to do conceptually the same thing on different types of data, without changing the name of the function, resulting in more efficiency.
  - **Operator Overloading**—Allows the programmer to give special meanings to an operator. This technique permits the programmer to add two user-defined variables with the same syntax that are applied to basic data types, thus improving extensibility and efficiency.

- **Abstraction**—Abstraction means to perceive an entity in a system, or context, from a particular perspective. A programmer takes out (hides) unnecessary details and only focuses on aspects that are necessary to that context or system under consideration. This removes distractions and complexity by only showing applicable desired information.
- **Encapsulation**—The binding of a class, consisting of its data, functions, subtask modules and properties, to shield it from outside interference and misuse. OOP takes advantage of application states and methods. Maintenance and testing times are reduced because the information is siloed, or shielded, from other data, therefore, reducing complexity and exposure.

Figure 1.18 below shows a high-level comparison between POP and OOP languages.

Figure 1.18—POP and OOP Language Comparison

Procedural-Oriented Programming (POP)	Object-Oriented Programming (OOP) (nonprocedural)
Uses algorithms	Uses data
Procedural (detailing **how** to achieve it)	Declarative (detailing **what** to achieve)
Programs are divided into functions	Programs are divided into objects
Most of the functions share global data	Data structures characterize objects
Data moves from function to function	Data is encapsulated and cannot be accessed by outside functions
Top-down approach	Bottom-up approach
Requires less memory	Requires more memory
No access specifiers	Access specifiers are private, public and protected.
Less secure	More secure
No overloading	Follows operator and function overloading
Examples: C, FORTRAN, COBOL, BASIC	**Examples:** JAVA, Python, C++, SQL

Java, Python and C++ are multi-paradigm languages and support both procedural and object-oriented programming.

1.2.18 Influential Software Enterprises

There are varying competing sources about who coined the term software first; recorded history documents Alan Turing as the first person to come up with a theory for software in 1935, which led to the two academic fields of computer science and software engineering.

Computer science is the branch of science and technology that is concerned with methods and techniques relating to data processing performed by automatic means and is theoretical. Software engineering is the application of a systematic, disciplined, quantifiable approach to the development, operation and maintenance of software, i.e., the application of engineering to software.

The following enterprises made notable contributions to software development:

International Business Machines Corporation

International Business Machines Corporation, better known as IBM incorporated, was founded in 1911 and was a merger of three smaller enterprises: International Time Recording Company, Computing Scale Company, and the Tabulating Machine Company to form the Computing-Tabulating-Recording Company (C-T-R), renamed IBM in 1917.[24] IBM, a pioneer in software development, made notable contributions, some of which are listed here:

[24] IBM, "1917," IBM Archives, www.ibm.com/ibm/history/history/year_1917.html

- Perhaps the earliest icon of the Information Age was a simple punched card produced by IBM, commonly known as the IBM card. Measuring just 7 3/8 inches by 3 1/4 inches, the piece of smooth stock paper has held nearly all the world's known information, for just under half a century. The first IBM card had 22 columns and eight punch positions; then 24 columns and 10 positions in 1900. In 1928, IBM introduced a new version of the punched card with rectangular holes and 80 columns.[25] Each hole punched correlated to 1 bit of data, or piece of information.
- The first mainframe computer was named the IBM 360 and can be thought of as servers. Mainframes are large, powerful computers usually used by businesses.
- IBM 305 Random Access Method of Accounting and Control (RAMAC) (also known as random access magnetic disk drive) was the first computer disk storage system. Such machines became the industry basic storage medium for transaction processing.
- IBM 701 and the Symbolic Assembler was the first commercial scientific computer. Nathaniel Rochester wrote the first symbolic assembler, which allowed programs to be written in short, readable commands, rather than pure numbers or punch codes.[26]
- In 1957, the first programming language was created. IBM introduced FORTRAN (FORmula TRANslation), a computer language based on algebra, grammar and syntax rules. It became one of the most widely used computer languages for technical work.[27]
- In 2004, IBM introduced the DS8000 series in dual two processor or dual four processor configurations, which can address over 96 petabytes of data, more than 4,500 times the amount of information found in the Library of Congress.[28]

There are many more innovations launched by IBM worth exploring, such as their notable work with the NASA spaceflight missions, namely Apollo, or the very latest artificial intelligence Watson computer. Watson is a question-answering computer system capable of answering questions posed in natural language, helps predict and shape future outcomes, and automates complex processes.

IBM is one company that was at the forefront of technology and is still making a difference 109 years later. IBM innovation crosses many domains, including healthcare, accounting and military, for hardware and software.

Microsoft

Microsoft was founded on April 4, 1975, by Bill Gates and Paul Allen in Albuquerque, New Mexico. Its current best-selling products are used by the majority of people using a myriad of devices. As the world's largest producer of software for personal computers, Microsoft offerings were initially geared towards the end user, an individual personal computer, hence the term microcomputer.

Notable Microsoft additions to software are as follows:

- In 1975, Microsoft created BASIC, a programming language interpreter for the Altair 8800 microcomputer.[29] Soon after, most home computers were using BASIC, such as the Commodore personal computer, IBM PC and Apple II.
- OS/2 operating system for IBM computers.
- Windows operating system as known today began as Windows 1.0 in 1985, improving the user interface with computers.

[25] *Ibid.*
[26] *Ibid.*
[27] IBM, "Chronological History of IBM: 1950s," IBM Archives, www.ibm.com/ibm/history/history/decade_1950.html
[28] *Ibid.*
[29] Microsoft, "Microsoft Fast Facts: 1975," 9 May 2009, https://news.microsoft.com/2000/05/09/microsoft-fast-facts-1975/

- Microsoft Office, a suite of productivity software comprised of PowerPoint for presentations, Word for writing and assembling documents, Excel for spreadsheet and advanced calculations, and many others.
- Xbox, a video gaming console and a line of entertainment games, music, and video.
- Bing, a search engine.
- Microsoft Azure, a cloud services platform.

Microsoft Beginners' All-purpose Symbolic Instruction Code (BASIC) programming language interpreter allowed hobbyists to write program code, but more so, it expanded functionality by adding a floating-point (first versions supported integer math only), expanding computing from 32-bit to 40-bit. The programming language interpreter, also known as assembler or compiler, translates the program commands to language that the computer can understand.

The decimal point of a number, known more commonly in computers as binary point, can float; that is, it can be placed anywhere relative to the significant digits of the number. This position is indicated as the exponent component, and, thus, the floating-point representation can be thought of as scientific notation.

A pivotal moment for Microsoft, and technology in general, is *United States v. Microsoft Corporation* antitrust lawsuit in 2001. Microsoft was accused of creating a monopoly on software and primarily placing political pressure, and legal and technical restrictions on the abilities of personal computer manufacturers and end users to uninstall Internet Explorer and use other programs. Furthermore, it prevented any integrations with any other applications (e.g., Java). The defeat of Microsoft enabled the open market of today and removed a technological blockage primed for an explosion of innovations.

SAP

SAP was founded in Germany in 1972 by five former IBM engineers. SAP stands for *Systeme, Andwendungen, Produkte in der Datenverarbeitung*, which—translated to English—means Systems, Applications, Products in Data Processing.

SAP was the first to offer a unified solution. While other applications solved one business need, such as accounting or payroll, SAP offered one application to manage the needs of all common departments via a modularized approach. The software allowed businesses to choose only what they needed with the ability to scale and customize to activate the other integrated modules on an as-needed basis. Following are the SAP modules:

- Enterprise Resource Planning (ERP) for inventory control
- SAP Financial Accounting (FI)
- SAP Controlling (CO)
- SAP Sales and Distribution (SD)
- SAP Production Planning (PP)
- SAP Materials Management (MM)
- SAP Quality Management (QM)
- SAP Human Capital Management (HCM)

Intel

Founded in 1968, Intel is a microchip processor manufacturer whose products are present in most personal computers today.

They manufacture SRAM and DRAM, motherboard chipsets, network interface controllers and integrated circuits, flash memory, graphics chips, embedded processors, and other devices related to communications and computing.

Intel invented the x86 series of microprocessors in 1978, which was sold on the IBM personal computer (PC) and is still used today on a myriad of brand name computers.

The x86 central processing unit (CPU) primarily handles programmatic functions and provides services, such as memory addressing, software and hardware interrupt handling, data type, registers and input/output (I/O) management.

Intel, a pioneering hardware company, enabled the exponential growth of personal computer purchases by creating a hardware architecture suited to the ever-changing complexity of software applications. As software became more complex and included thousands of lines of codes, the memory allocation and management of running the slew of programs had to allot for the demanding data sizes and make it affordable and small enough for personal computers and other Internet of Things (IoT) devices.

Oracle

Oracle, founded in 1977, is the largest relational database management system (RDBMS) company in the world, with presence in 175 countries, offering a myriad of complementary services, such as servers and storage systems, business intelligence and development tools.

An RDBMS is a relational model of a collection of information that is organized in tables and stored on a computer system; it is the most widely accepted database model today.

Over the years, Oracle had various new database releases, starting with the first database release in 1979. Oracle also made a variety of acquisitions, including PeopleSoft, Siebel Systems, Portal Software Inc., Agile Software Corporation, Hyperion Corporation, BEA Systems, Sun Microsystems, MICRO Systems and NetSuite.

Product highlights include:

- **2000**—Oracle E-Business Suite 11i and Technology Network (OTN) Xchange are introduced
- **2001**—Oracle 9i Database introduces Oracle RAC
- **2003**—Oracle Database 10G is launched as the first grid computing database available for the enterprise
- **2006**—Oracle announces Unbreakable Linux, a Linux distribution packaged freely distributed by Oracle, available partially under the GNU General Public License.[30]

Apple

Apple is a PC and software manufacturer and is arguably the biggest rival of Microsoft. Apple was founded by Steve Wozniak and Steve Jobs in 1976. Their claim to fame and main differentiators are twofold:

- Apple macOS is based on Unix, while Microsoft operating system software is based on MS-DOS.
- macOS is touted for its graphic interface, unique user experience and attractive hardware designs.

Unix is known to be a more robust operating system that offers finer levels of control and security; it is preferred by advanced users that can perform all functions via the command line instead of a user interface, which nontechnical end users prefer. Moreover, it is extensible and offers open source options on the market to extend functionality.

Apple, one of the most renowned innovators, is responsible for several advancements:

[30] Timelines, "Timeline of Oracle Corporation," 30 May 2018, https://timelines.issarice.com/wiki/Timeline_of_Oracle_Corporation

- In 2003, iTunes changed the music industry, and, in 2008, the application (App) store revolutionized software application distribution and sharing across devices. It created an entirely new digital purchasing and product/media distribution ecosystems by allowing users to easily buy music, books and applications online and have them show up on their computer and portable devices.
- In 2007, Apple introduced the touch screen and converted the phone hardware-driven dial pad to a software-driven interface with the launch of the iPhone and its mobile operating system named iOS. iOS created a whole new development language (and certifications) specifically for mobile devices, expanding developers' skill sets and technological advancements for mobile and wearable devices.
- Apple Watch introduced wearable technology into the mainstream. It was well-received and offered health management applications, such as pulse, sleep tracking, GPS exercise and other health-related features that were not accessible before in this form.

Adobe

Adobe and PDF (portable document format) is well-known for helping the world transition from paper to digital storage. It allows users to read and notate a myriad of document types (.doc, .ppt, etc.) without requiring the application in which the documents were created. It further expands the portability of application outputs that can be read by a PDF reader regardless of operating system or applications dependency.

VMware

VMware is a company that specializes in the area of virtualization and workstation optimization. It allows individuals and companies to slice a single PC or server into multiple virtualized, or emulated, computers, allowing the host computer and the emulated computer to have different operating systems and the emulated computer to have virtualized hardware.

Software has steadily evolved through the years from programs placed on discs to nearly immediate access through online activation and use, and, thanks to cloud technology, without having to install the software on the local PC. Customers are able to benefit from the innovations almost immediately with the ability to work faster and more efficiently.

1.3 Machine Code Concepts

1.3.1 Introduction

The term coding in software engineering is the process of expressing a computer program in a programming language. The transformation of logic and data from design specifications into a programming language creates an application that solves a problem.

There are coding standards and written procedures describing coding style, rules, keywords, etc. that govern the use of individual constructs provided by the programming language. The coding standards of naming, formatting, commenting and documentation prevent programming errors, control complexity and promote understandability of the source code. There are coding standard differences and similarities among the various programming languages.

A programming language is the way that a human communicates with a computer. A computer-specific language is needed because a computer works only with its machine language (bits and bytes that are expressed in 0s and 1s). This language is slow and cumbersome for humans to use. Therefore, programmers write applications in a computer programming language, which then gets converted into machine language for the computer to process.

There are many computer languages, and they have been evolving from complex machine language into a more natural way of writing. Programming languages have evolved to support the kind of application that they intended to solve and the kind of approach used in the design.

Each phase of a programming language evolution can be thought of as a generation.

1.3.2 Central Processing Unit

The central processing unit (CPU) of a computer includes the circuits controlling the interpretation of program instructions and their execution, and controls the entire computer. The CPU sends and receives data through input-output channels, retrieves data and programs from memory, and conducts mathematical and logical functions of a program.

Every computer contains a CPU. It loads the program to the memory and processes the sequence of instructions—code—into machine language for the computer to execute. The instruction set consists of native data types, addressing modes, instructions, registers, memory architecture, interrupt and exception handling, and external input/output (I/O).

The main function of the CPU is to receive input from a peripheral such as mouse, keyboard, monitor, touch screen, scanner, printer or computer program, and interpret what it needs (**figure 1.19**). The CPU does all this mechanically—without thinking about or understanding what it does—therefore, the program it executes must be complete in all details and unambiguous, because the CPU can do nothing but execute the program exactly as it is written.

A CPU contains an arithmetic logic unit (ALU) that carries out operations, such as addition and subtraction. The ALU contains small memory units, called registers, capable of holding a single number. Computer memory capacity is measured in terms of how many bits it can store, hence the type of numbering system it uses.

Figure 1.19—CPU Schematic

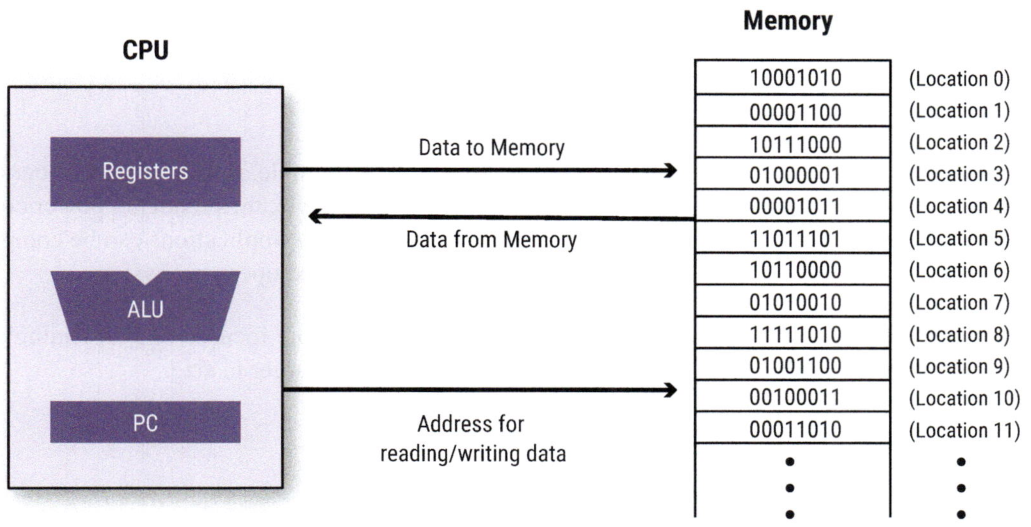

Source: Eck, D.; *Introduction to Programming Using Java, Eighth Edition, Version 8.1*, July 2019, Section 1.1, April 2020

1.3.3 Instruction Set

Machine-language instructions are patterns of bits with specific patterns corresponding to specific commands to the machine. The instruction set is part of the computer architecture known as the instruction set architecture (ISA). ISA is the boundary between the software and the hardware. An instruction set provides commands to the computer's CPU and specifies the computer functionality, as follows:

- The CPU memory storage mechanisms and how they are accessed (i.e., similar to house and street address)
- The operations that the CPU supports
- How the programmer/compiler communicates with programs

The more instructions (i.e., more features) that a CPU has, the more circuits that are required to implement the instructions.

In computing, an operation code (opcode), also known as instruction machine code, or opstring, is the portion of a machine-language instruction that specifies the operation to be performed.

1.3.4 CISC and RISC Instruction Set Architecture Differences

Intel's x86 family of ISA first began with 16-bit processing in 1978 and now offers a 64-bit chipset. A 16-bit processor means that the CPU can process data and memory addresses[31] that are represented by 16 bits. 64-bit

[31] Beal, V.; "Address," Webopedia, https://www.webopedia.com/definitions/address/

supports vastly larger amounts of virtual memory and physical memory than is possible on its 32-bit and 16-bit predecessors, allowing programs to store larger amounts of data in memory.

The 32-bit and 64-bit notation for a processor terms reference how a processor (CPU) handles data. A 32-bit architecture allows the arithmetic and logic unit (ALU), or digital circuit, to perform 32-bit integer arithmetic and logical operations.

Intel processors fit into an ISA family called complex instruction set computing (CISC). The Intel competitor ARM uses reduced instruction set computing (RISC) architecture.

RISC processors focus on keeping the number of instructions to as few as possible, reducing the number of circuits and reducing the size, while also keeping those instructions as simple as possible, thus reducing power consumption. Those benefits come at a cost of increasing execution time, and not all software applications can be compiled to install and run natively on devices with ARM RISC processors and the intended operating system.

Unlike RISC computers, the instructions available on a CISC processor are more focused on performing complex tasks with large amounts of flexibility, but require more power and are likely larger in size.

1.3.5 Machine Code Language

Machine code, the first generation language, is a strictly numerical language—a machine language instruction is just a sequence of zeros and ones. These binary instructions are intended to run as fast as possible, and may be regarded as the lowest-level representation of a compiled or assembled computer program or as a primitive and hardware-dependent programming language.

Binary in mathematics and computers, refers to a base 2 numerical notation and consists of two values: 0 and 1, which can also notate an electrical circuit on and off, respectively.

Each 0 or 1 is called a bit. Each particular sequence of 0s and 1s encodes a particular instruction. The data that the computer manipulates is also encoded as binary numbers. In modern computers, each memory location holds a byte, which is a sequence of eight bits.

Figure 1.20 shows the base and characters (numbers and letters) for the common computer numbering systems and the decimal system.

Figure 1.20—Number Representation Techniques

System	Base	Characters
Binary	2	0,1
Octal	8	0,1,2,3,4,5,6,7
Decimal	10	0,1,2,3,4,5,6,7,8,9
Hexadecimal	16	0,1,2,3,4,5,6,7,8,9,A,B,C,D,E,F

1.3.6 ASCII and Unicode

A computer understands only numeric values. Prior to 1963, computer manufacturers had over 60 different ways of representing characters in computers. Machines could not communicate with one another. This problem was becoming increasingly evident as companies began networking computers.

The International Organization for Standardization (ISO), the world's largest developer of voluntary international standards, released ISO/IEC 2022, specifying the structure of 8-bit codes and 7-bit codes, which provide for the coding of character sets.[32] This created a standard that everyone can follow to enable networking and a common programming language encoding foundation.

The American Standard Code for Information Interchange (ASCII) characters are used in programming, data conversions, text files, graphic arts and in emails. Programmers can use ASCII to represent calculations on characters.

ASCII is a 7-bit code (2^7), so it has 128 defined characters. ASCII code consists of 95 printable characters, including letters, punctuation marks and numbers, and 33 nonprintable control characters that send commands to the computer or a printer.[33] When there is a piece of text, ASCII converts each character to a 7-bit binary number (**figure 1.21**). This set of numbers is easier to store into the computer memory. Assigning a number to a character is called encoding.

Figure 1.21—ASCII Conversion Table

Character	Binary Code	Character	Binary Code	Character	Binary Code	Character	Binary Code
A	01000001	U	01010101	o	01101111	)	00101001
B	01000010	V	01010110	p	01110000	*	00101010
C	01000011	W	01010111	q	01110001	+	00101011
D	01000100	X	01011000	r	01110010	,	00101100
E	01000101	Y	01011001	s	01110011	-	00101101
F	01000110	Z	01011010	t	01110100	.	00101110
G	01000111	a	01100001	u	01110101	/	00101111
H	01001000	b	01100010	v	01110110	0	00110000
I	01001001	c	01100011	w	01110111	1	00110001
J	01001010	d	01100100	x	01111000	2	00110010
K	01001011	e	01100101	y	01111001	3	00110011
L	01001100	f	01100110	z	01111010	4	00110100
M	01001101	g	01100111	!	00100001	5	00110101
N	01001110	h	01101000	"	00100010	6	00110110
O	01001111	i	01101001	#	00100011	7	00110111
P	01010000	j	01101010	$	00100100	8	00111000
Q	01010001	k	01101011	%	00100101	9	00111001
R	01010010	l	01101100	&	00100110	?	00111111
S	01010011	m	01101101	'	00100111	@	01000000
T	01010100	n	01101110	(	00101000	_	01011111

Unicode is a universal coding system designed to be used with different language scripts. Unicode has three types of available encoding: UTF-8, UTF-16 and UTF-32. UTF-8 uses 8 bits per character, UTF-16 uses 16 bits per character and UTF-32 uses 32 bits per character. In UTF-8, the first 128 characters are the ASCII characters. Therefore, ASCII is valid in UTF-8. Usually, Unicode is used in internationalization and localization of computer software. This standard is also used in operating systems, extensible markup language (XML), .NET framework and programming languages, such as Java.

32 International Organization for Standardization (ISO), *ISO/IEC 2022:1994 Information technology – Character code structure and extension techniques*, www.iso.org/standard/22747.html

33 IONOS, "ASCII – American Standard Code for Information Interchange," 8 May 2019, www.ionos.com/digitalguide/server/know-how/ascii-codes-overview-of-all-characters-on-the-ascii-table/

ASCII can only represent 128 characters; it includes lowercase letters (a-z), uppercase letters (A-Z), digits (0-9), symbols (e.g., punctuation marks) and control characters. Unicode represents letters of English, Arabic, Greek, etc., mathematical symbols, historical scripts, emoji—a wider range of characters when compared to ASCII.

1.3.7 Assembly Machine Language

Prior to the second generation of languages, programs were created using a machine code. When programming languages advanced beyond machine code, an intermediary had to be used, a translator of sorts. Assembly machine language is this intermediary language. Assembly language uses mnemonics to present the instructions to be translated later into machine language by an assembly program. The assembler translates source code files written in assembly language into their machine language equivalents, as shown in **figure 1.22**.

Figure 1.22—Assembly Machine Language Translation Process

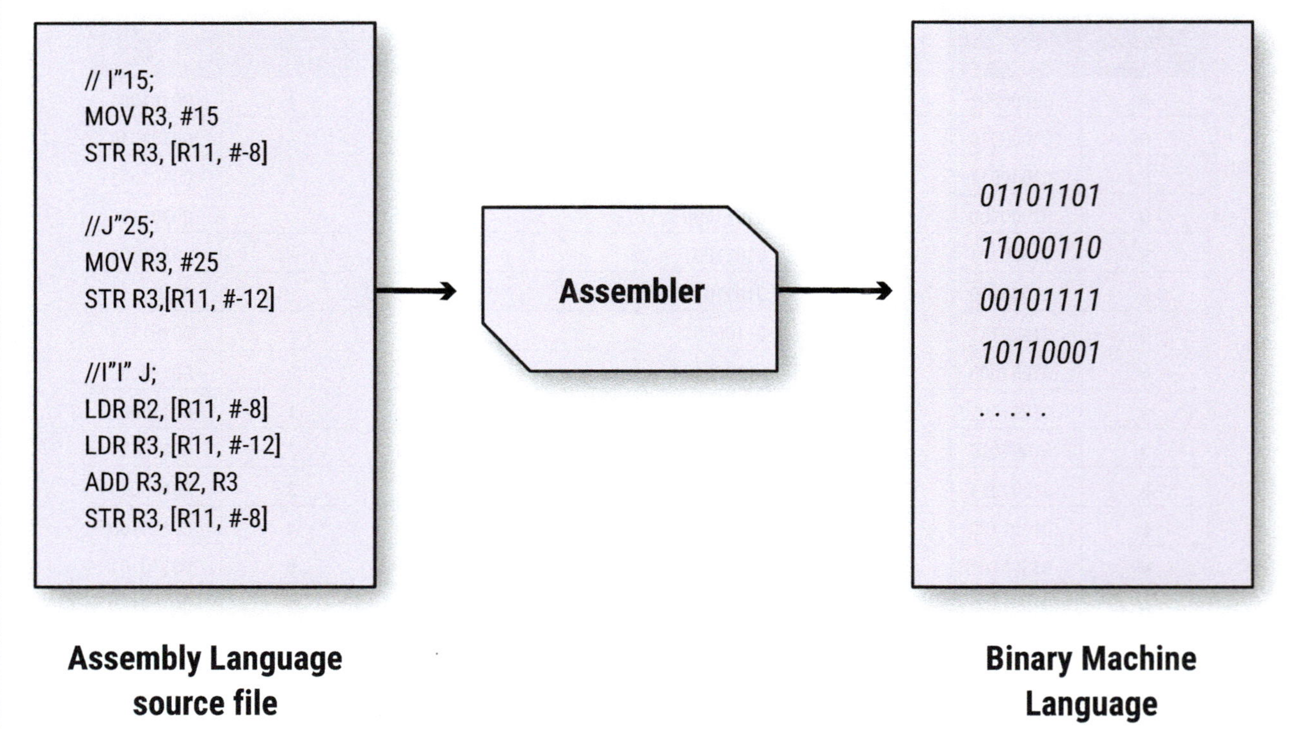

Source: Sims, G.; "Assembly language and machine code—Gary explains," Android Authority, March 2016, www.androidauthority.com/assembly-language-and-machine-code-678230/

Using assembly language removed the need to communicate in binary. Assembly language is easier for humans to remember than the binary codes.

Mnemonics is a symbol or name that is used to define or specify a computing function for humans to remember easily. An example of this is opcode 00—written as "STOP" in assembly language—which stops program execution. A program written in assembly language is compiled and executed via an assembler to the processor.

Because assembly language is the closest to the hardware (i.e., second generation), each assembler is highly dependent on the hardware type. A different assembler is needed depending on the assembly language and hardware used.

1.3.8 Interpreters and Compilers

Procedural languages, also called third-generation, high-level languages (HLL), such as Pascal, FORTRAN, COBOL, Basic, and C, use compilers. The coded program, called a source program, has to be translated through a compilation step, before the program can be executed.

A compiler translates programs that are expressed in a high-level language into their machine language equivalents in one step, unlike assemblers or interpreters, which translate and execute one instruction statement at the time of execution (known as runtime).

The interpreter must be resident in the computer each time a source-code file, which is written in an interpreted language, is executed.

Compiler and interpreter comparison is shown in **figure 1.23**.

Figure 1.23—Comparison of Compiler and Interpreter

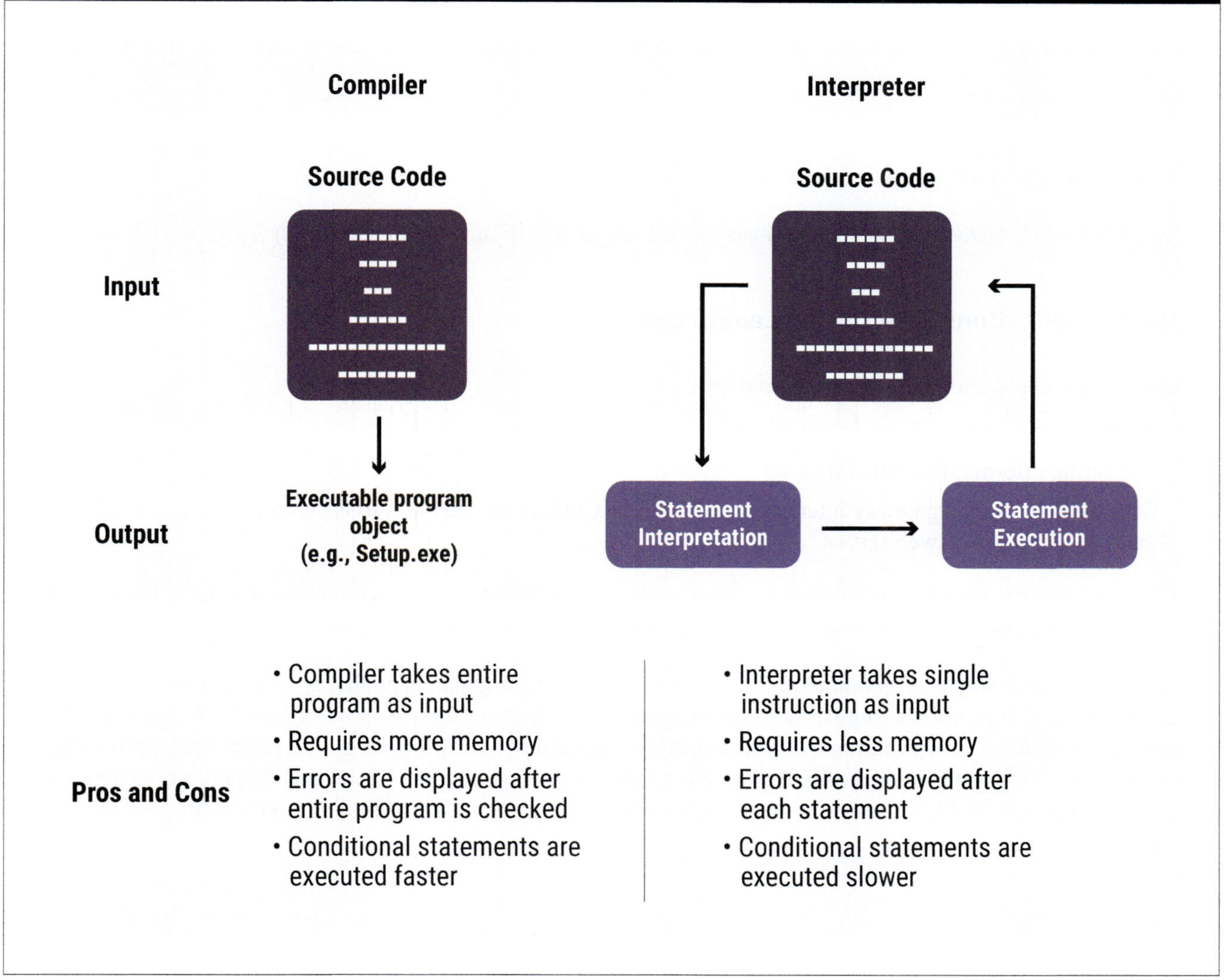

1.3.9 Scripting

All scripting languages are programming languages, although much smaller (in feature set and size) and are usually limited to a specific task. More importantly, scripts require a host program in which to run.

The theoretical difference between the high-level programming languages and scripting languages is that, normally, a high-level program, such as C, needs to be compiled before running; whereas, a scripting language, like JavaScript, Python, Perl or PHP, typically does not need to be compiled. The use of a compiler makes the application hardware independent. Scripting languages use interpreters and, therefore, are dependent on the hardware (much like the assembly language), which must be taken into consideration.

Furthermore, some scripting languages (such as Java Script and Python) can be compiled (into bytecode—not machine code) and then used on any platform via a virtual machine.

Bytecode, as shown in the following example, is a binary representation of an executable program that is designed to be executed by a virtual machine, rather than by a dedicated hardware CPU. The bytecode program now becomes portable and hardware independent, because it is processed by software. Bytecode uses a compiler that is often named Just in Time (JIT) to signify it produces the bytecode during runtime, and not at the end of the function.

Example:

Python uncompiled line of code: a = b.c()

Python compiled (bytecode): load 0 (b); load_str 'c'; get_attr; call_function 0; store 1 (a)

1.3.10 Applications of Scripting Languages

Following are applications of scripting languages:[34]

- Automating certain tasks in a program
- Extracting information from a data set
- Creation of common gateway interface (CGI) scripts. CGI scripts enable web browsers to make use of programs that are running on a web server.

Due to steadfast advancement in modern hardware computation abilities and in coding practices, the distinction between scripting languages and high-level programming languages is getting hard to recognize.[35]

PowerShell is an example of a command-line interface (CLI) scripting language that is used on the Windows operating system to perform computer administration tasks, such as configuration management. A command-line interface allows users to type commands to the operating system in a scripting language, rather than use the graphical user interface. Bash is the command-line interface scripting language used on UNIX-based operating systems, like Linux, Ubuntu and HP AUX. Bash is a very secure approach to work with any kind of UNIX secure code deployment.

[34] GeeksforGeeks, "What's the difference between Scripting and Programming Languages?," 21 January 2019, www.geeksforgeeks.org/whats-the-difference-between-scripting-and-programming-languages/

[35] *Ibid.*

1.3.11 JAVA and .NET Virtual Machines

Programming languages have advanced to include simpler and faster ways of programming, gradually striving to become platform-agnostic, portable, scalable and easier for others to leverage. Programming languages are increasingly allowing a program to be used in more systems than initially intended.

Moreover, software development skills are now portable too—if Java is a developer's expertise, his or her Java program can be compiled and used in web development where C++ experience was previously needed.

A virtual machine (VM) is a virtual emulation of a computer operating system. There are two different implementations of this kind of system abstraction:

- System virtual machines are created in such a way to virtualize all hardware, network connectivity, operating system, memory allocation and the like to exist as fully functional computers and are usually created as an alternative for using a physical machine. VM environments are created on actual hardware (the host) that is located on or off premises. A software hypervisor separates the host machine resources from the host hardware so the resources can be used by the virtual machines (guests). In this way, multiple environments can run on the same machine simultaneously (e.g., running a Windows operating system on a Mac via a VM.) This is especially helpful when test, staging, and development environments are set up with sample data. After the virtual environment is used, it can be reset to its original state and used again—this is a space, time and cost savings option.
- The common language runtime (CLR) and Java virtual machine (JVM) are both procedural virtual machines. Instead of mimicking an entire operating system, they are specifically designed to run an individual process or program without being dependent on the platform environment.

1.3.12 Portability of Code

The main difference between Java and .NET virtual machines is that Java Enterprise Edition (JEE), can work on any operating system, while .NET works only on various versions of Windows, but the .NET CLR is language neutral. Java takes advantage of the Java Virtual Machine (JVM), which follows the concept that code should run on any device and under any operating system, thus making itself a portable and platform-independent programming tool. JVM was originally Java language specific.

The .NET platform is the unified environment provided by Microsoft and thus has proprietary translation to Microsoft Intermediate Language (MSIL). It provides seamless horizontal scalability where most tools and IDEs can be found within the Microsoft Developer Network and complete ecosystem of integration (**figure 1.24**). Although CLR can be used on Windows operating systems only, CoreCLR is an open source, cross-platform subset of the .NET platform that can be extended to operating systems such as Linux or Mac's OSX.

Figure 1.24—Virtual Machine Portability

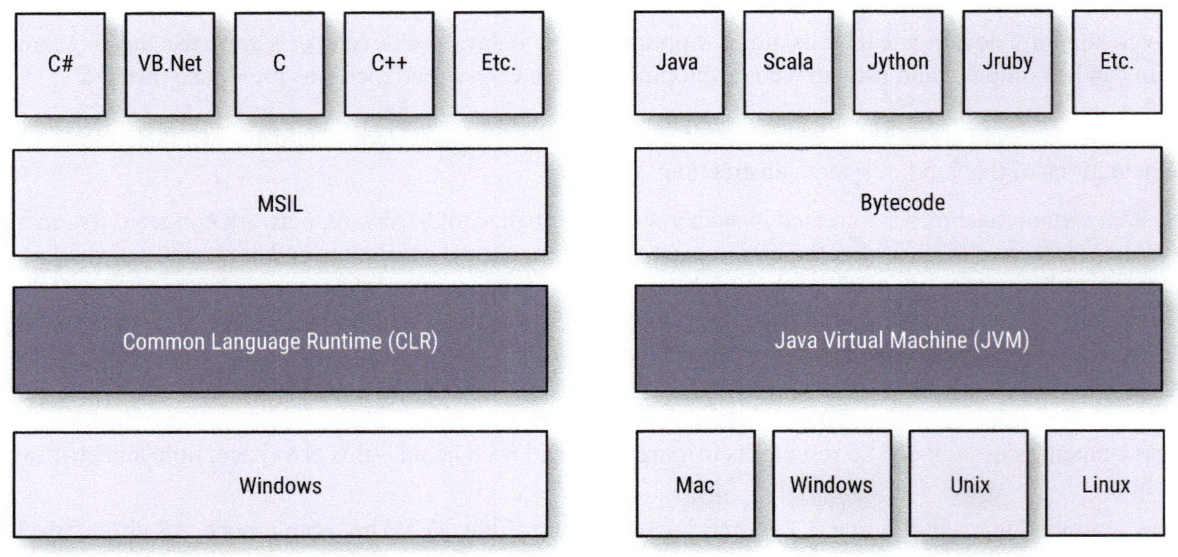

Source: Soroker, T.; "CLR vs. JVM: How the Battle Between C# and Java Extends to the VM-Level," OverOps, May 2018, https://40c98d22-a-62cb3a1a-s-sites.googlegroups.com/site/sitinnova/asix_m01_uf1/nf1-virtualization/processvirutalmachines.png?attachauth=ANoY7coOdAJyu2a41AUjo-UJFNppTGJ9YSqpHsoG8UtB0Ks54i23JAR_P0DciGOm_Md1YM3o32t0KeLBt1kUedhe3FHPFBTZjkMU6ILEqDrDJ95rDR7uL82lCkn4fqylzj-iaoimd9Kj7_iRug0ZlMd5l53f8OhpO5WRmHycoUoZi8BNEskMgDi6PkI28YvXXiyRTaJL21-jxYVZvPF4EIAvOrKrD2ZVvqwsA2j6flpozS_mEMU8BqUicDf58JThPbMEbNB8990A7JGjvN_LdVCTuBWOcOPx1Q%3D%3D&attredirects=0

1.4 High-Level Languages

1.4.1 Overview

Software programs can be written in many languages. As discussed in section 1.3, machine code is the lowest-level language comprises numerical values only. Machine code is highly dependent on the hardware and difficult and cumbersome to create because it is the furthest away from natural human language.

Each computer language evolved separately, driven by the adaptation to new standards, hardware capabilities and innovation. The following sections describe several of the most widely used computer languages. Languages can be sorted into two main categories: low-level languages and high-level languages. A language overview is shown in **figure 1.25**.

Figure 1.25—Overview of Software Languages

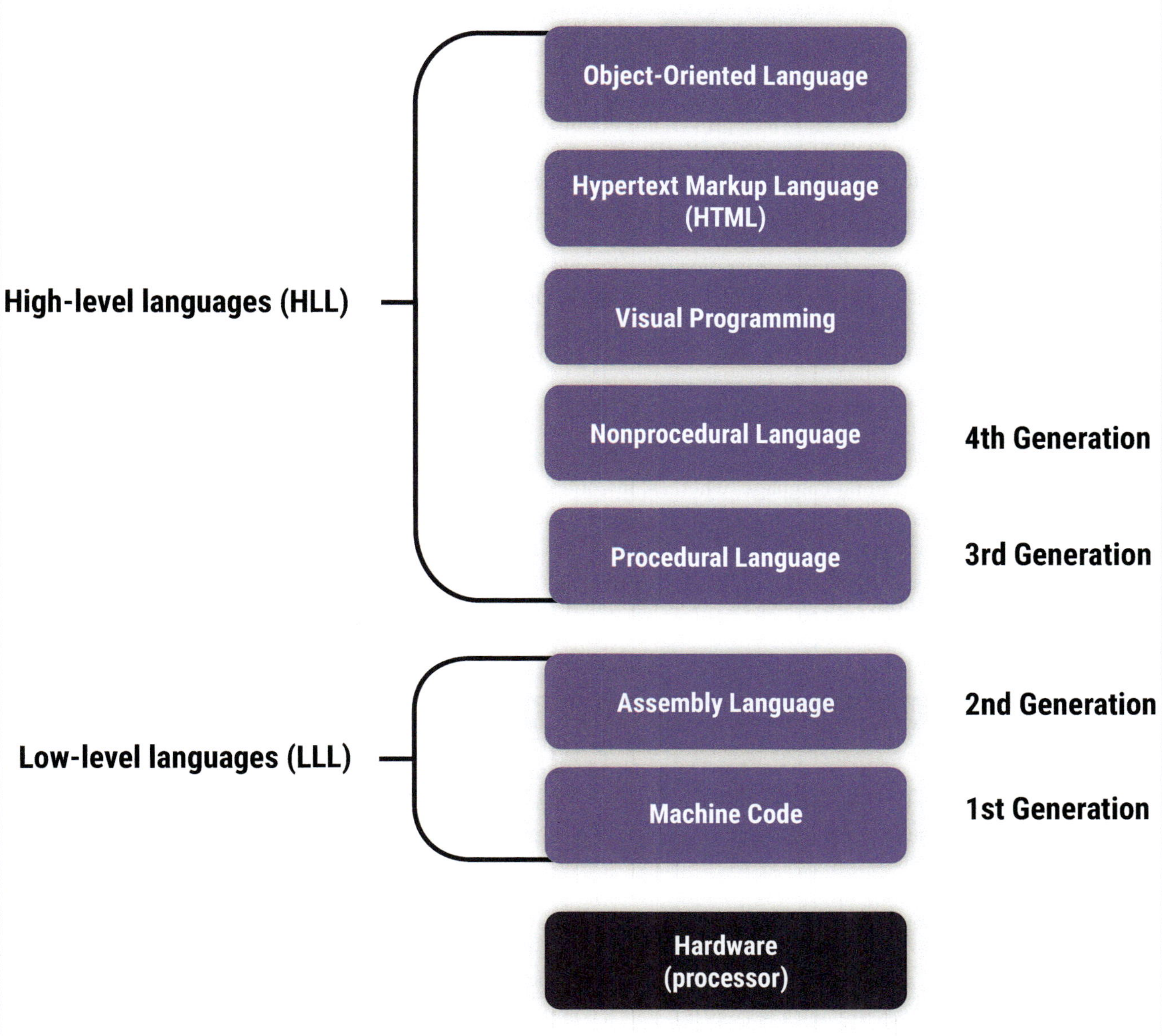

Higher-level languages are easier to read, write and maintain because they are closer in lexicon to natural language. For the computer languages under discussion, they are classified into the following categories: procedural and nonprocedural, visual, graphic, functional, object-oriented and scripting languages.

High-level languages are portable and are not machine dependent. The programs created using these languages can work across numerous CPU families and can support a range of data types. Moreover, these languages can be translated into several target machine languages, allow symbolic naming of operations and addresses, provide features designed to facilitate expression of data structures and program logic, and usually result in several machine instructions for each program statement, thus greatly expanding the use and scalability of the application.

One disadvantage is they are slower than low-level languages, given they are higher, i.e., further away from machine language.

1.4.2 Procedural and Nonprocedural

Procedural and nonprocedural languages are the computation models for specifying most programming today. The major difference between these computational models is that the procedural language is *command-driven* whereas nonprocedural language is *function oriented.*

Procedural language (PL), also known as procedural oriented programming (POP) is third-generation and includes languages such as Pascal, FORTRAN, COBOL, Basic and C. The program has to be translated through either a compiler or an interpreter as discussed in section 1.3.8. As the name suggests, PL requires a set of step-by-step commands completed in order and relies on well-organized procedures, functions or subroutines.

Nonprocedural languages are fourth-generation, also known as 4GL, and are used for predefined functions in applications for databases, report generators and queries. Nonprocedural languages are declarative languages—users concentrate on defining the input and output rather than the program steps required. An example is Structured Query Language (SQL), which is used for querying a database for specific data.

1.4.3 C and C++

C is a procedural programming language, whereas, in contrast, C++ is an object-oriented programming language (OOP) that uses objects, classes polymorphism and inheritance overloading. These are not present in C language; C is a subset of the C++ language, while C++ is a superset of C.

Figure 1.26 further describes the differences between the two high-level languages.

Figure 1.26—Comparison of C and C++ Programming Languages

C	C++
Created in 1969	Created in 1979
Subset of features in C++, cannot run C++ code	Can run most of C code
Procedural (means no support for polymorphism, encapsulation or inheritance)	Procedural and object-oriented programming are supported.
Function-driven	Object-driven
Data and function are separated.	Data and function are encapsulated together in the form of an object. The class is the outline of the object.
Data is not hidden.	Encapsulation hides the data.
Multiple global variables declaration is allowed.	Multiple global variables declaration is not allowed.
No function and operator overloading.	Function and operator overloading capable.
Functions cannot be defined inside structures.	Functions can be defined inside structures.
No exception handling.	Exception handling is supported.
Declaring variables must be done initially at the start.	Declaring variables can be done anywhere.

Global variables are declared outside of functions; these variables can be accessed, read or used by any function in the program. Global variables remove redundancy and allow for declaration upfront in some cases (see **figure 1.26**). However, the variable name must be unique, and, if it is a long and complex program, keeping track of the status and name of that variable (to avoid duplicates) may be cumbersome.

Local variables are declared inside a function, and can only be used inside that specific function. When using local variables isolated to functions, the same variable name may be used in other functions with different data types (e.g., *varA* in **function A** may be an integer, but *varA* in **function B** may be a character).

1.4.4 JAVA

Java, created by Sun Microsystems, refers to both a programming language and a platform for development of applications that work on multiple operating systems, such as Windows, or Mac OS or Linux. Java programming language and platform consists of many software components that work together to provide a cross-platform environment. These components include Java Development Kit (JDK), Java Runtime Environment (JRE), and Java Virtual Machine (JVM).

Previously, it was well integrated in the World Wide Web interactive functionality using applets—small applications installed as part of the browser to play a video or other interaction. Today, a Java application programming interface (API) framework is part of the Android platform, available on some mobile devices. An API allows two applications to communicate and transfer input to each other. Programmers use these Java APIs while creating Android apps.

1.4.5 HTML

Hypertext Markup Language (HTML) is used for writing World Wide Web Internet applications, such as websites. It comprises many client-side text pages that are *marked* to denote text, color, size, font, etc. Cascading style sheets (CSSs) are also used, to create web-page structure and further design elements.

Considering that the Internet is a server-client configuration, many websites use active server page (ASP) HTML pages that include one or more scripts (small embedded programs, such as JavaScripts) that are processed on a web server before the page is sent to the user via a browser.

WordPress is an open-source, widely used application for web development. Small businesses and individuals can create websites in as little as a few minutes due to its graphical interface and drag-and-drop functionality (**figure 1.27**).

Moreover, the text editing, which can be done via HTML by skilled developers, can be hidden and replaced by a WYSIWYG (What You See Is What You Get) editor, which works and looks like any other text editing application, such as Microsoft Word.

Figure 1.27—Example WordPress Platform (Usage at 37.8% Across the Web as of July 2020)

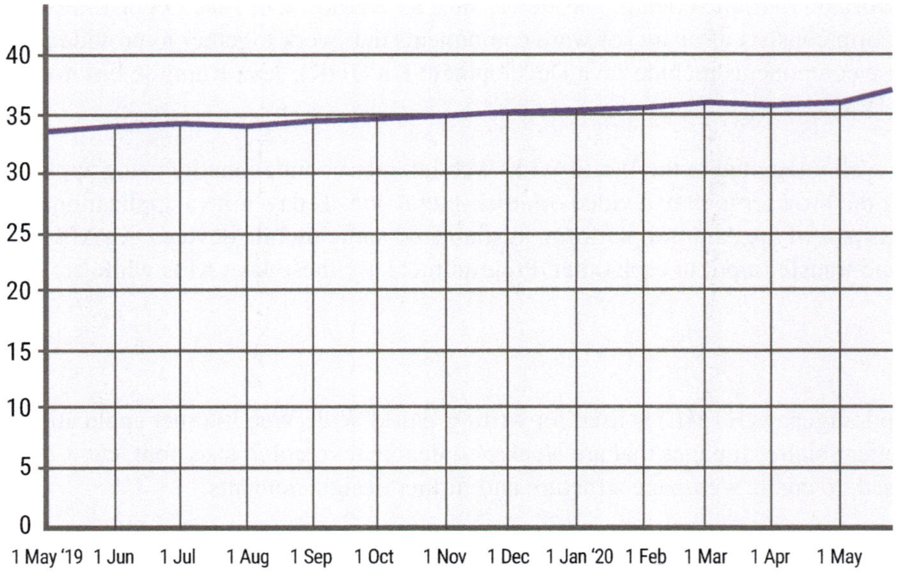

Usage of WordPress, 26 May 2020, W3Techs.com

Source: W3techs.com, "Usage statistics and market share of WordPress," https://w3techs.com/technologies/details/cm-wordpress

1.4.6 Integrated Development Environment (IDE)

An integrated development environment (IDE) is a software application that programmers use to assist them with application development. In general, an IDE is an application with a graphical user interface (GUI) combined with an integrated environment that aids a developer in building software applications with all the required tools at hand. An IDE normally consists of at least a source code editor, build automation tools and a debugger for finding errors. A developer can write code, analyze, debug, test, collaborate, compile and deploy directly from the IDE.

There are several integrated development environments:

- Microsoft Visual Studio products support application development for C++, C# and VisualBasic.NET languages (**figure 1.28**).
- Oracle NetBeans can be used for developing Java applications, C/C++, Ruby, HTML5 and PHP.
- Idle and Atom are two examples for Python development.

There are many other IDEs on the market, and some are open source applications offered at no cost. Most IDEs offer a rich user interface with extensible add-ons, wizards, libraries of example code and much more, assisting the developer with writing quality applications.

Figure 1.28—Microsoft Visual Studio Integrated Development Environment

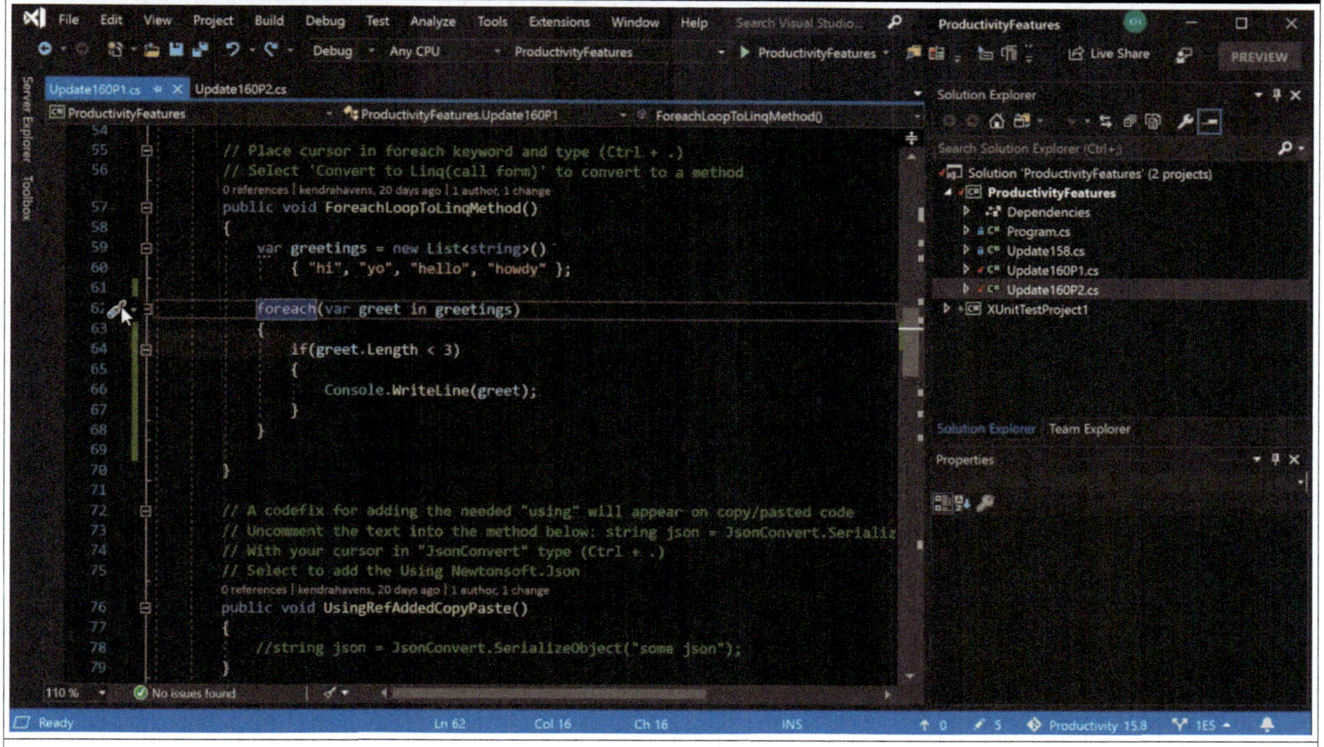

Source: Microsoft, "Visual Studio 2019," https://visualstudio.microsoft.com/vs/

1.4.7 Python

Python is an interpreted, object-oriented, high-level programming language with dynamic semantics.[36] Its high-level build in data structures, combined with dynamic typing and dynamic binding, make it very attractive for rapid application development (RAD), and for use as a scripting or glue language to connect existing components together. Python's simple, easy-to-learn syntax emphasizes readability and, therefore, reduces the cost of program maintenance. Python supports modules and packages, which encourages program modularity and code reuse. The Python interpreter and the extensive standard library are available in source or binary form without charge for all major platforms, and can be freely distributed.

Often, programmers fall in love with Python because of the increased productivity it provides. Because there is no compilation step, the edit-test-debug cycle is incredibly fast. Debugging Python programs is easy: a bug or bad input never causes a segmentation fault. Instead, when the interpreter discovers an error, it raises an exception. When the program does not catch the exception, the interpreter prints a stack trace. A source-level debugger allows inspection of local and global variables, evaluation of arbitrary expressions, setting breakpoints, stepping through the code a line at a time, and so on. The debugger is written in Python itself, testifying to the Python introspective power. On the other hand, often the quickest way to debug a program is to add a few print statements to the source; the fast edit-test-debug cycle makes this simple approach very effective.

[36] Python, "What is Python? Executive Summary," www.python.org/doc/essays/blurb/

1.4.8 .NET Suite and Visual C++

.NET, provided by Microsoft, is an open source, free, cross-platform, development application for building many types of programs. It allows developers to code multiple languages, and use libraries and editors in one platform. This greatly improves efficiency and reduces the learning curve necessary for multiple platforms or tools when creating complex applications, whether they are used for web, desktop, gaming or the like.

Moreover, the .NET framework and family of products is a foundational application prerequisite that enables the code to run natively on any compatible Windows OS. .NET supports a myriad of language families (it uses C#, pronounced C-sharp): C, C++, Java, JavaScript and Visual Basic. The .Net Core is a cross-platform .NET implementation for websites, servers and console apps. The latest, Xamarin/Mono, is a .NET implementation for running apps on all the major mobile operating systems.

The Microsoft ecosystem of applications also includes a standard API for all Windows-based applications in addition to a robust set of programming libraries.[37]

Visual programming languages (also called block-based coding) allow programmers to manipulate program elements graphically by using a mouse and icons to design the application, rather than specify elements and instructions textually via lines of code. In these languages, the user selects values from a dropdown, and then drags and drops elements of code or fields to design the front-end, visible user interface. The code is assembled on the back-end, where it is hidden from view. This approach is used for students, initial software application design, graphical user interfaces or 3D application designs.

[37] Microsoft, "What is .NET?," https://dotnet.microsoft.com/learn/dotnet/what-is-dotnet

1.5 Chapter 1 Knowledge Check

REVIEW QUESTIONS

1. Choose the **BEST** answer. What role does CPU play in a computer as it relates to software?
   A. Allocates memory needed for the program
   B. Mechanically executes a set of instructions exactly as they are written
   C. Converts programs to machine code
   D. Performs arithmetical and logical calculations

2. BIOS and drivers are an example of:
   A. Utility software
   B. Application software
   C. System Software
   D. Firmware

3. Choose the **BEST** answer. Binary in mathematics and computers refers to a ___________ numerical notation.
   A. Base 2
   B. Octal
   C. Decimal
   D. Hexadecimal

4. Which of the following is the **BEST** definition for software?
   A. A set of one or more interrelated programs capable of being executed
   B. A collection of programs, procedures, documents and rules that direct computer operation via lines of instructions
   C. The American National Standards Institute (ANSI) defines it as a set of one or more interrelated modules capable of being executed
   D. A process by which user needs are translated into a software product

5. Which of the following companies is credited with creating the first interpreter language?
   A. IBM
   B. Microsoft
   C. Intel
   D. SAP

Answers on page 60

Chapter 1 ANSWER KEY

Review Questions

1. A. Allocates memory needed for the program
   B. Mechanically executes a set of instructions exactly as they are written. Refer to page 44.
   C. Converts programs to machine code
   D. Performs arithmetical and logical calculations

2. A. Utility software
   B. Application software
   C. System Software
   D. Firmware. Refer to page 19.

3. **A. Base 2. Refer to page 46.**
   B. Octal
   C. Decimal
   D. Hexadecimal

4. A. A set of one or more interrelated programs capable of being executed
   B. A collection of programs, procedures, documents and rules that direct computer operation via lines of instructions. Refer to page 16.
   C. The American National Standards Institute (ANSI) defines it as a set of one or more interrelated modules capable of being executed
   D. A process by which user needs are translated into a software product

5. A. IBM
   B. Microsoft. Refer to page 40.
   C. Intel
   D. SAP

SOFTWARE DEVELOPMENT FUNDAMENTALS
An ISACA Certificate

Chapter 2:

Programming

Programming

Programming

2.1 Learning Objectives

After completing this chapter, learners will be able to:

1. Describe input types (standard input [STDI], user input, process input, APIs).
2. Identify the basic structure of a program.
3. Describe data processing.
4. Recall data types (integers, characters, strings, etc.).
5. Compare synchronous vs. asynchronous processing.
6. Recall Boolean logic operators.
7. Explain the flow of control structures (if/else, while, do/while, etc.).
8. Identify data structures and data storage types.
9. Describe output types (peripherals, standard outputs [STDO]).

2.2 Input

The behavior of software programs is dependent on input. A computer program interacts and communicates with the outside world to get the data (input) needed for its programs and to communicate the results (output) of its data manipulations. The process part of the cycle is sometimes referred to as a black box by software testers and end users, because the code and the interaction between the program and the CPU and memory are not seen (**figure 2.1**).

Figure 2.1—Overview of the IPO (Input, Process, Output) Cycle

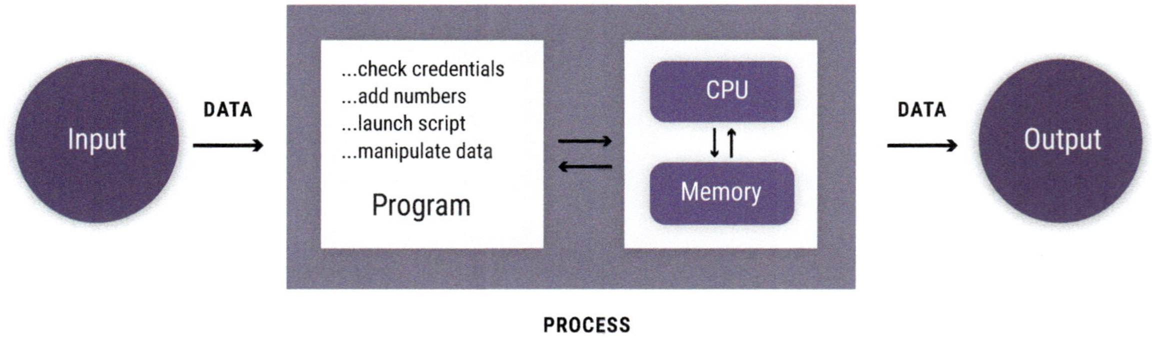

2.2.1 Input Data

Input data must first be loaded to memory before it can be manipulated. The program allocates the memory necessary for each data input based on the data type set by the programmer in advance. Entering data can be accomplished via the following methods.

- Mouse
- Keyboard
- Scanner

- Microphone
- Joystick
- Webcam
- Trackball
- Digital camera
- Digital pen
- Touch screen

Additional data inputs are the newest Internet of Things (IoT) wearable devices, such as watches that receive blood pressure, heart rate and other biometric data.

2.2.2 Process Input

An application programming interface (API) is an example of program-to-program input. An API is a way for an application to interact with certain systems, applications, routines, script, protocols, library, etc. to either send or receive input without human interaction.

A good API makes it easier to develop a program by providing all the building blocks related to functional characteristics of an operating system that applications need to specify. For example, when interfacing with the operating system (e.g., Microsoft Windows or different versions of UNIX), a programmer utilizes APIs in developing applications that can operate effectively and efficiently on the platform chosen.

A good API serves the intended purpose of its design, whether it is integrating separate applications to use the same single sign on, or collecting and transferring information between two systems.

2.2.3 Data Types

Data types or simply type is an attribute of data which tells the compiler or interpreter how the program intends to use the data for processing and how much memory to allocate for the input.

Data types are often grouped in two main categories, as shown in **figure 2.2**. Primitive, or fundamental, data types are basic to the language. Derived data types are defined by the programmer and are aggregates of fundamental data types.

Figure 2.2—Data Types: Primitive and Derived

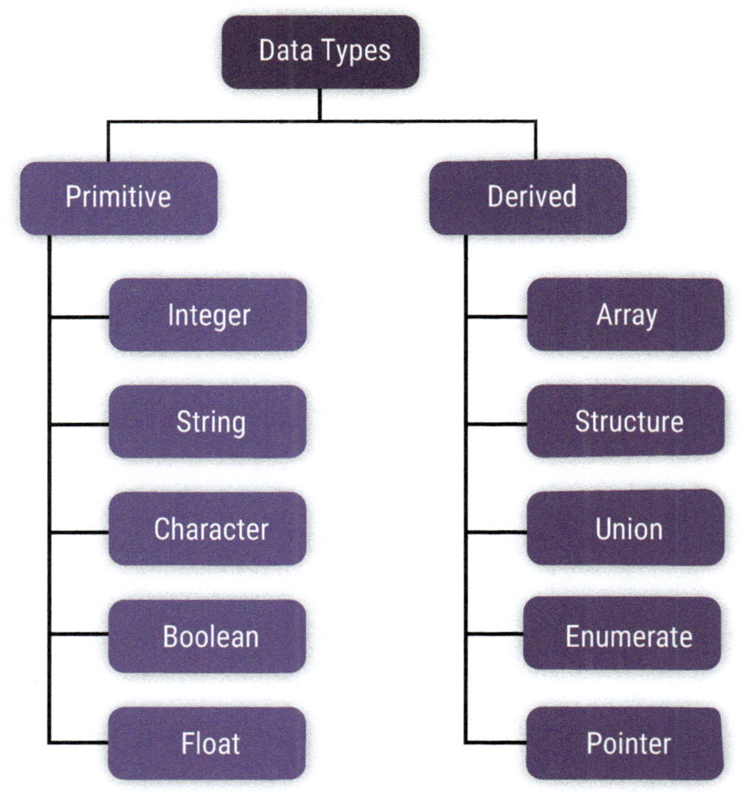

Source: Adapted from Meritcampus, "2. Data Types," http://java.meritcampus.com/core-java-topics/java-data-types

Following are descriptions of the primitive data types:

- **Integer**—Represents some range of mathematical integers. Integers are whole numbers and are commonly represented in a computer as a group of binary digits (bits of 1s and 0s). For example: 21, 4, 0, and −2048 are integers, while 9.75, 5+1/2, and √2 are not.
- **String**—A series or a set of characters (i.e., char) as a single piece of data, most often notated with double quotes. The string may be constant (fixed, unchanging) or a variable (changing). A string variable holds a character string and is given a name by the programmer. For example: "Hello World" is a string, and "F" is a character string; The following is a string variable example: If a program asks the user for their first name, and the user inputs John, then the program outputs "Hello John", the program variable <name> was set equal to "John" while processing the rest of the instruction set.
- **Character**—A single-digit string, unlike a string data type that can allot for many more characters. For example: "F" when the program asks users if they are male, they may respond with an F for False, at which point the program will continue asking only female-relate questions.
- **Boolean**—Allows one of two possible values: 1 or 0, which correlate to TRUE or FALSE or ON or OFF, respectively.
- **Float**—A floating-point number, which has a decimal place. Float is more precise than an integer, stores up to seven digits and occupies four bytes of memory (32 bits). If more than seven digits are present, the value is rounded off. For example: 19.95 or 3.1415. Double is twice as precise (hence the name), storing up to 15 digits and occupying eight bytes of memory (64 bits).

Following are descriptions of the derived data types:

- **Array**—A collection of memory locations, all storing data of the same type (e.g., a collection of integers). For example: int[ ] numberArray = {1, 5, 9, 13, 17};
- **Enumerated**—Allows the programmer to invent his/her own data type and decide what value the variables of this data type may carry. The enum (i.e., enumerator) term is commonly used in mathematics, software engineering and computer science to refer to a listing of all the elements of a set. Generally, it is a complete, ordered listing of all the items in a collection. For example: public enum Days { Monday, Tuesday, Wednesday, Thursday, Friday, Saturday, Sunday}
- **Structure**—A term used to describe a grouping of different types into a single type. For example: struct Car { char make[30]; char model[30]; int vin[17]}
- **Union**—A term used to describe differing data types sharing one memory location. For example: union Car { char make[30]; char model[30]; int vin[17]}
- **Pointer**—References a location in memory

Data types can be further subcategorized into three types (**figure 2.3**).

Figure 2.3—Data Types

Category	Examples	Description
Simple data types	Integer, character	Example int x = 7 Variable x is set to type int, and assigned a value of 7.
Structured data types	Array, strings	Example (using C++) of an array, which is a data structure that stores a fixed number of values for a variable: int myArray [4] = {10, 20, 30, 40}; The array, named myArray, can hold 4 integer values. Array values are accessed by their index number, which always starts at 0 (10 is at index 0 in the example).

User-defined data types	Enumerate, structure, unions	Example (using C++):
		struct tenant
		{
		int tenantid;
		char name[20];
		char address[50];
		};
		Notice that name and address are set to use specific memory size.
		The structure tenant, has three members (tenantid, name, and address) with two different data types (int and char).

If a value entered does not comply with the expected input for a program, it could result in an error or an unhandled exception, e.g., entering "Hello World" where a numerical value is expected.

2.3 Processing

The best way to learn programming is to practice writing programs and get familiar with an IDE (see chapter 1). This section explores the fundamental structure of a program using mostly C++ as the programming language. The following is an example.

```
1  // my test program utilizing C++
2  #include <iostream>
3
4  int main()
5  {
6     std: :cout << "Hello World!";
7  }
```

Note: Normally, some parts of code will be highlighted in color. The colors may differ based on the IDE tool used or developers' preference (dark or light themes, etc.). An example color scheme is to use green to signify comments and blue to notate keywords (e.g., int in this case to signify integer) and preprocessor instructions (e.g., #include). Colors are not visible in this guide.

The numbers on the left are part of the code-writing application to easily reference a line of code and are not part of the program instruction set.

- **Line 1**—Two slashes indicate the line is a comment inserted by the developer, and it will have no effect on the behavior of the program.
- **Line 2**—A hash tag (#) is a directive read and interpreted by the preprocessor. The hash tag instructions are interpreted before the compilation of the program itself begins. In this example, the line #include <iostream> incorporates a C++ code standard used for standard input and output operations, such as writing the output of this program (Hello World!) to the screen.
- **Line 3**—Blank lines do not affect the behavior of the program and assist in reading the code.
- **Line 4**—int main() is the start of a function. A function is a grouping of code instructions which are given a name: main is the name of the function—the group of code statements that follow. The function declaration includes: data type (int), a name (main) and a pair of parentheses (()) which may hold optional parameters. A function name of main is a special function in C++ and is the first function to run, regardless of where it is located in the program.
- **Lines 5 and 7**—Open and close braces signify the beginning and end of the function.
- **Line 6**—This instruction statement has four parts:
  - **std: :cout**—Signifies a standard character output device, such as the monitor. The program will print/display output to this device. To avoid typing "std: :" each time there is character output, the following namespace can be added: using namespace std;
  - <<—This is the insertion operator, which indicates that the elements after it will be inserted into the standard character output.
  - **"Hello World!"**—Everything in double quotes is a string that is output to the monitor
  - ;—The semicolon ends all C++ statements, much like a period in a sentence

2.3.1 Synchronous and Asynchronous Processing

Synchronous and asynchronous are processing terms used to describe *when* a program executes the input instruction set. Synchronous processing executes at specific time intervals or when another process on which it is dependent has completed. It can block other processing and create bottlenecks. Asynchronous processing is independent of the timing requirement and can process multiple instruction sets concurrently. The execution of different actions is done at the same time without waiting for another.

A common user experience is having multiple programs running when the user opens a document and starts typing, but no letters appear in the document. Shortly after, all the letters appear at once in the exact order they were typed. This is a synchronous processing example of a common bottleneck, due to the OS being slowed down, causing other applications to be delayed, as the OS is responsible for resources allocation, such as memory, processing, etc.

2.3.2 Boolean Logic and Operators

Boolean logic operators AND, OR and NOT are expressions used to control the data set that is returned. The program instruction is to influence the results based on varying criteria.

The operators AND, OR and NOT are used to either broaden or narrow the returned search results (**figure 2.4**). This is especially helpful when searching a relational database with many tables for specific records.

Figure 2.4—Boolean Logic Operators

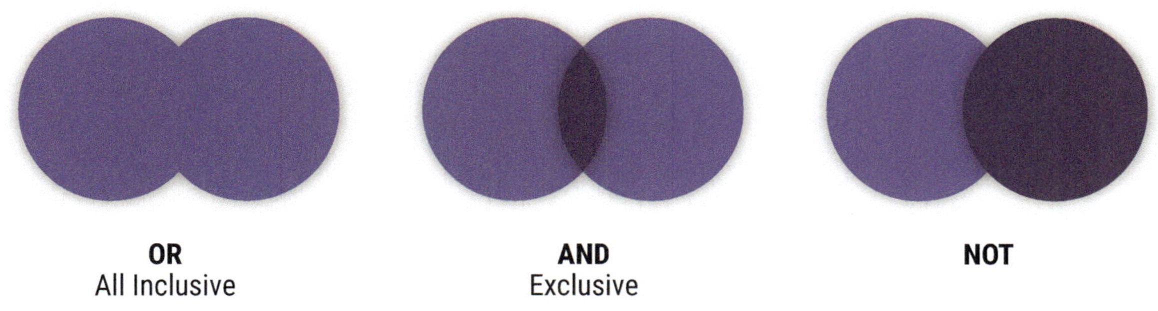

Example: Boolean logic is used every time a search engine is used to structure a query on the backend. For example, a search for *ISACA certification* may or may not translate to ISACA *and* certification and the results can include the combination of all three words. Performing a few preliminary searches will reveal how the backend query is used and the search terms can then be tailored accordingly.

Following are the results returned from using a popular search engine to perform some tests (**figure 2.5**); the goal is to find all sites that have the ISACA-provided certifications only.

Figure 2.5—Example Search Engine Results

Search String	Result
isaca certification	1,510,000 results
ISACA CERTIFICATION	1,410,000 results
isaca and certification	1,460,000 results
isaca or certification	1,470,000 results
isaca	3,650,000 results
certification	788,000,000 results
"isaca certification"	109,000 results
"isaca" "certification"	1,250,000 results
"isacacertification"	300 results

Which search term syntax is the best one to use to get the most relevant results?

In some search engines, the keyword AND is an implied operator. In others, such as relational databases, the Boolean operator AND is *not* implied and so the results may not be as relevant. An additional difference is that relational database SQL queries can be set to be case sensitive: certification is different from CERTIFICATION.

Using double quotes ("") around data communicates a complete string to the program and it has to be processed exactly, with the words precisely as they are with the space—or not.

Searching a database using SQL is also a very common example of the Boolean operators, especially in a relational database that is constructed of many tables that join based on a primary or foreign key (secondary key). The tables have relationships that further describe the data attributes.

A primary key uniquely identifies each record in a table. Primary keys must contain unique values and cannot contain NULL (empty) values. A table can only have one primary key and can consist of single or multiple columns (fields). Primary keys can be likened to a person's social security number and are never reused; they can be automatically generated by a database (also referred to as GUID for global unique identifier) each time a record is created or manually set by a database administrator.

A foreign key is used to link two tables together. A foreign key is a field or several fields that reference the unique primary key in another table. For example, a renter will have a renter ID in the *Tenants* table that is a primary key for that table that references renter's apartment, number 20, with a foreign key of apt ID that is a primary for that apartment in the *Apartments* table.

Using the property management example:

- One property building may have one apartment number 20—this is a one-to-one relationship
- One apartment, number 20, may have many tenants—this is a one-to-many relationship

The records are further refined with active/inactive notation to keep the integrity of the relationships (e.g., an apartment may have had multiple tenants, but only one renter is active at once).

In **figure 2.6**, the relation type is also depicted as 1 and ∞ (infinity symbol) to denote a 1 to many relationship—one customer may have multiple orders to infinity and each "order/customer" combination is a unique, separate record.

Figure 2.6—Illustration of One and Many Relationship Types

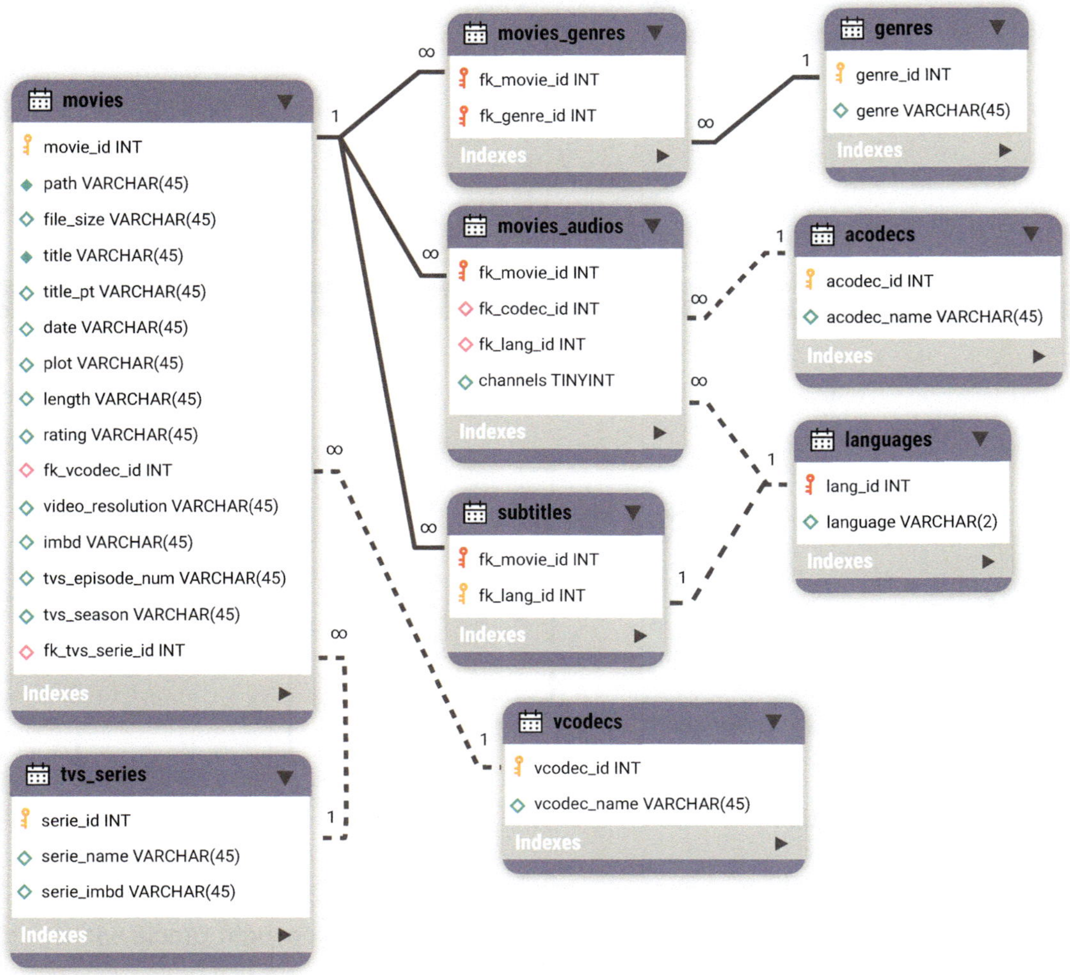

Source: StackOverflow, "MySQL Modeling: Relationship normalization & ring issue," https://stackoverflow.com/questions/29990685/mysql-modeling-relationship-normalization-ring-issue

2.3.3 Control Structures

Flow of control structures is the order in which individual instructions, statements or function calls are executed or evaluated within a program (**figure 2.7**). Each programming language has a reserved set of words, also called keywords, that are reserved as a predefined instruction set and have an inbuilt meaning.

Notice the indentations. Indentations in most programming languages are only used to ease the strain of reading code—the computer ignores them. However, in some languages, such as Python scripting language, indentations are critical for the code to process correctly. Python code indentations replace braces, semicolons and other structural notations available in other languages. In the following code snippet example, there would be no way to know if the second Print statement belongs to the if statement or the f function with this indentation structure.

Example:

```
def f(x):
  if x == 5:
    print("x is five")
    print("who am I?")
```

Figure 2.7—Basic Types of Control Structures

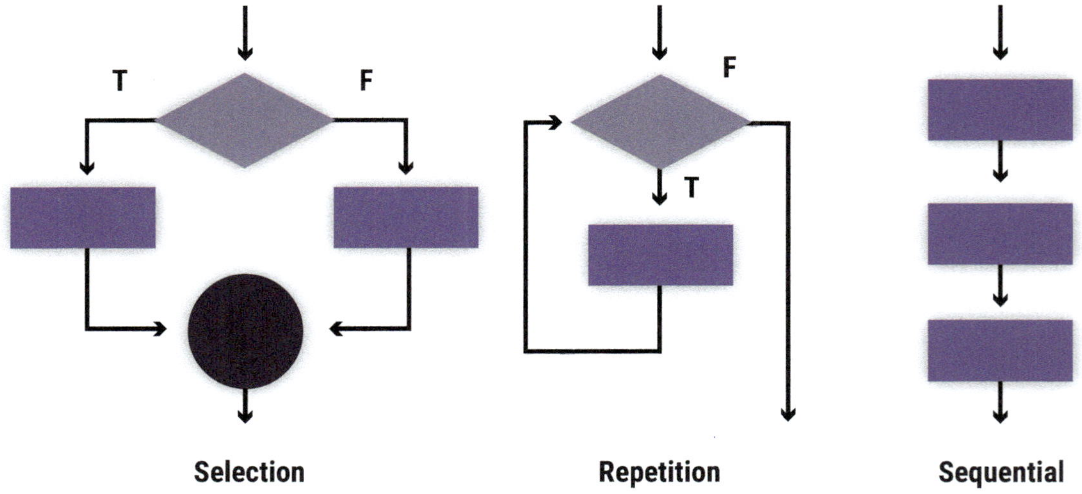

Source: Oracle, "PL/SQL User's Guide and Reference Release 8.0," figure 3-1 Control Structures, https://docs.oracle.com/cd/A64702_01/doc/server.805/a58236.pdf. Copyright Oracle and its affiliates. Used with permission.

- **Selection**—Used for decisions and branching—choosing between two or more alternative paths. In C++, the types of selection statements are if and if/else. Some other language also use if/then selection statements.[38]
- **Repetition**—Used for looping, i.e., reiterating a piece of code multiple times in a row. In C++, there are three types of loops (note that the keyword may vary from language to language, but the principal use is the same). In C++, the types of repetitive statements are while, for, do while.
- **Sequential**—Consecutive execution of code statements—one line after another. This is the default mode.

The following sections describe the reserved words with which one should be familiar in the software development domain.

2.3.4 If Decision Making

When the reserved word *if* is found, the interpreter expects an expression (condition) to check if the expression is true or false. The action statement that follows instructs the computer to do something if the expression is true. If the expression is false, the computer ignores the action statement and reads the next *else* line and performs the *else* action statement.

```
if (expression)
    action statement
else
    action statement
```

[38] "Control Structures – Intro, Selection," www.cs.fsu.edu/~myers/c++/notes/control1.html

The following example is checking for *one condition* only:

Example (using C++)	Code

Example (using C++)	#include <iostream>
	int main ()
	{
	 int grade=68; // initializing variable
	 if (grade >= 68) // is the grade greater or equal to 68?
	   std::cout << "Passing!"; // prints the word Passing!
	}

When variables are first declared, they do not have a value (e.g., int grade;). A variable is assigned a value, as in the previous example (e.g., int grade=68;).

If checking for *two conditions*, add the word **else**.

Example (using C++)	Code

Example (using C++)	#include <iostream>
	int main ()
	{
	 int grade=60; // initializing variable with 60 grade
	 if (grade >= 68) // is the grade greater or equal to 68?
	   std::cout << "Passing!";
	 else
	   std::cout<<"Failing"; // prints the word Failing
	}

Else is not required if there is only one condition. Moreover, in some languages, like Visual Basic .NET, if/then decision statements are used, as shown in the following example.

Example	Code

Example (using Visual Basic .NET)	Imports System
	public class MainClass
	  Shared Sub Main()
	    Dim grade As Integer = 55 // declaring and initializing the variable grade with 55
	    If grade >= 68 Then // if grade is greater than or equal to 68, then go to the following line
	      Console.WriteLine("You passed")
	    Else
	      Console.WriteLine("You failed") // prints you failed
	    End If
	  End Sub
	End Class

Figure 2.8 shows a conditional flow for the if operator.

Figure 2.8—Conditional Flow for IF Operator

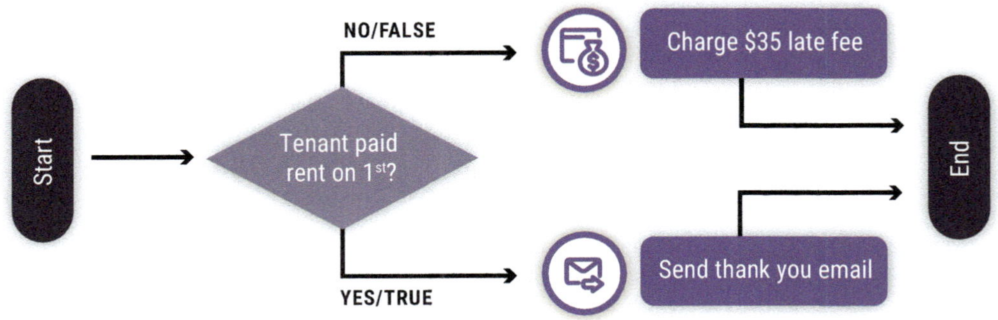

If tenant paid rent on the 1st (=yes/true), *then* send thank you email, *otherwise* charge $35 late fee.

2.3.5 While Loops

A **while loop** is used to repeat a specific block of code continuously until a condition is met. There are several ways to implement a while loop, depending on what we are trying to capture and whether the number of times the loops will be executed is known.

A **for loop** is used when it is known at compile time how many times this loop will execute. A **while** loop is normally used in a scenario where it is not known how many times a loop will actually execute at runtime.

```
while(condition)

{

  statement(s);                    }
```

Note: In some languages, like Visual Basic .Net, the statement ends with End While.

```
While (condition)

        [ statements]

        End While
```

Example:

i is set to equal 1 and is being increased by a number of iterations. Each time the loop runs, it adds +1 until **i** is equal to 8.

Example (using C++)

```
#include <iostream>
/* The loop continues to print the value of i
        until the given condition i<=7 returns false. */
using namespace std;  //add this to load standard library to use cout
int main ()
{
int i=1;
  while(i<=7) {
    cout<<"Value of variable i is: "<<i<<endl; i++;   //endl; instructs the output device to go to end the current line of output and start a new line
    }
}
```

The output is as follows:

Value of variable i is: 1

Value of variable i is: 2

Value of variable i is: 3

Value of variable i is: 4

Value of variable i is: 5

Value of variable i is: 6

Value of variable i is: 7

2.3.6 Infinite While Loops

These loops are great for capturing errors because the program errors may not be known and how many times an error may appear during runtime.

Example: **i** is set to equal 1, and I is being decreased by a number of iterations. Each time the loop runs, it removes -1 until **i** is equal to 7 (which will never be true). Because i will never equal 7, the loop will never stop.

Example (using C++)

```
#include <iostream>
/* This loop will never end (infinite loop). The value of i is consistently being decremented.
The value of i is 1 so the condition i<=7 would never return false. */
using namespace std;  //add this to load standard library to use cout
int main ()
{
int i=1;
 while(i<=7) {
   cout<<"Value of variable i is: "<<i<<endl; i–; // i is being decremented –
   }
}
```

The output is as follows:

Value of variable i is: 0

Value of variable i is: -1

Value of variable i is: -2 and so on…

2.3.7 Do While Loops

A **do-while loop** is used where the loop should execute at *least one time* before checking the condition to see if it is true.

do {

statement(s);

}

while(condition);

In languages like Visual Basic .Net, there are two options for constructing the do-while loop Do { While } condition, as shown in the following example.

Do { While } condition	**Do**
[statements]	[statements] **Loop** { **While** | **Until** } condition
Loop	This loop should execute at least one time.

Example: int (whole number) variable **num** is set to equal 2 and is being increased by a number of iterations. Each time the loop runs, it adds +1 until **num** is less than or equal to 7.

Example (using C++)	#include <iostream>
	using namespace std;
	int main ()
	{
	int num=2;
	do {
	cout<<"Value of num: "<<num<<endl;
	num++;
	}
	while (num<=7);
	return 0; //adding this line is good practice and 0 implies the program ended successfully

The previous example is very similar to the example for the while loop, but here, the program is being run first (notice the arrow and the int num=2 is outside of the brackets) without checking if it is true; so, instead of printing 1 as the first output, it will print 2 and only then check the condition.

The output is as follows:

Value of num: 2

Value of num: 3

Value of num: 4

Value of num: 5

Value of num: 6

Value of num: 7 At this point **num** is equal to 7

2.3.8 For Loops

A **for loop** is used where it is known at compile time how many times this loop will execute.

For (initialization; condition; propagation)

{

statement(s)

}

In some languages, like Visual Basic .NET, For/Next is used, as follows:

For counter [**As** datatype] = start **To end**

Next [counter]

Example: Using C++, the value of **i** is set to equal 10. The condition is that the loop only runs when **i**<20 (until **i** equals 19). Every iteration adds +1. It is going to run 10 times and exit.

Example (using C++)	`#include <iostream>`
	`using namespace std;`
	`int main ()`
	`{`
	`  for ( int i = 10;  i < 20;  i = i + 1 ) {`
	`    cout << "value of i: " << i << endl;`
	`  }`
	`  return 0;`
	`}`

The output is as follows:

Value of i: 10

Value of i: 11

Value of i: 12

Value of i: 13

Value of i: 14

Value of i: 15

Value of i: 16

Value of i: 17

Value of i: 18

Value of i: 19

2.3.9 Nested Loops

Nested loops are loops within loops. They comprise outer and inner loops, a hierarchal structure of inner loops providing data to the outer loop. The parent has the final data value. For every iteration of the outer loop, all the iterations of the inner loop are performed. Finally, a while loop can be inside a for loop or vice versa.

Nested loops can become very complex during the creation process and when debugging any errors. This is due to a possible logical error beginning in any of the inner loops and propagating the bad data to each dependent loop. One helpful mitigation when using nested loops is to indent the loops to various depths to the right. This can help distinguish the various loops.

Nested loops can be used when a set of data is received, and parsing (i.e., further analyzing) is used to find a subset of the initial data. **Figure 2.9** is a general graphic representation. Some programs allow up to 250 inner loops whereas others allow 75.

Note: In the following example, ! represents NOT, and % is the modulo operator, which divides and returns the remainder.

Example (using C++)	
	```cpp #include <iostream> using namespace std; int main () { int x, y; for(x = 2; x<53; x++) { // find prime numbers between 2-53 for(y = 2; y <= (x/y); y++) if(!(x%y)) break; // if factor is found, it is not prime if(y> (x/y)) cout << x << " is prime\n"; } return 0; } ```

The output is as follows:

2 is prime

3 is prime

5 is prime

7 is prime

11 is prime

13 is prime

17 is prime

19 is prime

23 is prime

29 is prime

31 is prime

37 is prime

41 is prime

43 is prime

47 is prime

The following example shows nested For/Each loops in Visual Basic .NET. The example also shows the indentations to distinguish the different loops.

Example (using Visual Basic.NET)	<pre>Module Module1
	  Sub Main()
	    Dim apartments() As String = {"Beach Front", _ ' The input array
	                "With Balcony", _
	                "Without Balcony", _
	                "Backyard"}
	    ' Loop over each element with For Each.
	    For Each apartments As String In Apartments
	      Console.WriteLine(Apartments)
	    Next
	      Console.WriteLine(). ' Blank line
	    ' Use For-loop
	    For index As Integer = 0 To Apartments.Length -1
	      Console.WriteLine(apartments(index))
	    Next
	  End Sub
	End Module</pre>

The output is as follows:

Beach Front

With Balcony

Without Balcony

Backyard

Beach Front

With Balcony

Without Balcony

Backyard

Figure 2.9—Nested Loop With Two Conditions

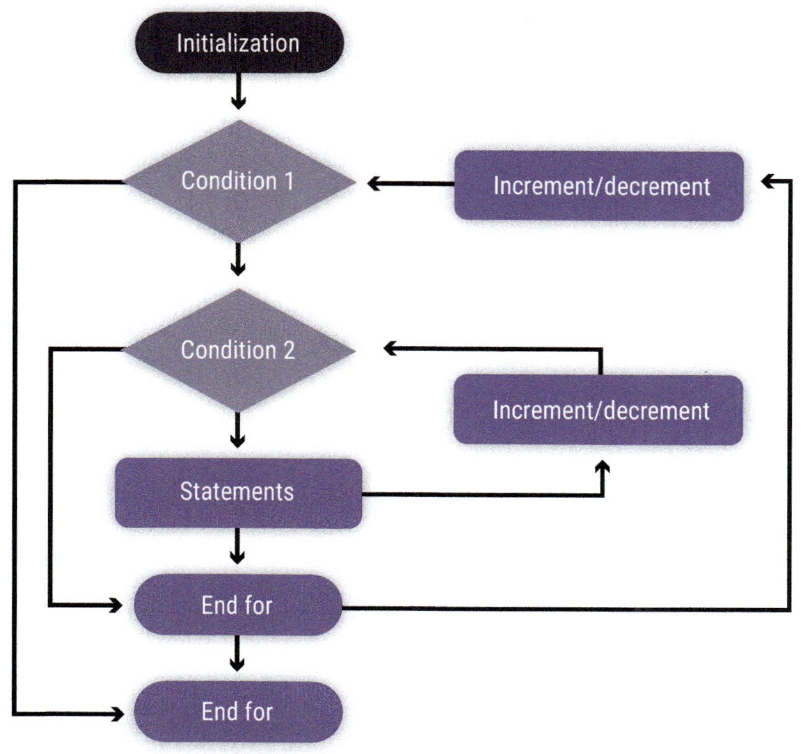

2.3.10 Data Structures

In software development, a data structure may be selected or designed to store data for the purpose of manipulating it with various algorithms and to organize code and information in a digital format. Each data structure contains data attributes (information) and relationships information between the data and functions that can be applied to the data. Using the correct data structures assists with proper organization and retrieval of stored information.

The following are the characteristics of data structures:

Static or dynamic: Static means fixed, unchanging. When the data is processed and compiled, static data fixed sizes, structures and memory locations remain the same. Dynamic data structures have memory allocations that can shrink or expand depending on the input and processing at compile time.

- **Homogeneous or nonhomogeneous**—This trait describes whether all data items are of the same type or of various types (e.g., integers or floats).
- **Linear or nonlinear**—Data can be ordered (i.e., chronological order; e.g., array) or nonordered (e.g., a graph).

Primitive data types, such as integers, floats, characters and Boolean, were discussed in section 2.2.3. Following is an overview of advanced data types:

- **Array**—Arrays are both static and dynamic and are used to store multiple data items as a grouping or a collection, using neighboring memory locations. The items stored this way are usually the same; hence, the position of each element can be calculated or retrieved easily.
- **Linked list**—A linked list stores a collection of items in a linear order. Each element, or node, in a linked list contains a data item and a reference, or link, to the next item in the list.

- **Stack**—A stack stores a collection of items in linear order; the order is last in first out (LIFO).
- **Queue**—A queue stores a collection of items similar to a stack; however, the operation order can only be first in first out (FIFO).
- **Tree**—As the name implies, a tree is an abstract, hierarchical node-based view that highlights the data relationship via nodes; also known as children.
- **Graph**—Graphs are useful for representing any network type of data, such as devices on a network. The data stored in graphs is not ordered (contrast with stacks) data; Graphs are made up of a predetermined set of nodes (e.g., vertices) and lines that connect them (e.g., edges).
- **Tries/trie**—It uses a visual graph that can store data strings graphically.
- **Hash table/hash map**—A hash table is a map, a schematic of where the data is stored. It uses an associative array that charts keys to values and a hash function to convert an index into an array of containers that contain the desired data item.

2.4 Output

2.4.1 Output to Peripherals

After the data have been processed, manipulated and analyzed, the application communicates the results to the end user (end user may be another application, computer or a person). Communicating results can be done via one or multiple methods (**figure 2.10**):

Figure 2.10—Communication Vehicles

Communication Method	Description
Visual cues	Devices, such as a monitor or printer, can display videos, images, errors and text.
Audible cues	Audible devices, such as headphones and speakers, are used to output music, videos and other notification sounds.
Background activities	Command, file or signal as input to another program.
Storage	Flat files from any program, such as .doc, .ppt, .jpg, .ini, and many more, can be saved either locally, externally or remotely for later retrieval or further analysis and input into another program. Reports can be created in formats such as extensible markup language (XML) or PDF and others. Various tools are available for the creation of reports as output from the developed application. These tools include Microsoft Power BI, Oracle Reports, Oracle Discoverer, SAP Crystal Reports, Oracle XML Publisher, SQL Server Reporting services, Tableau and Power BI.

2.4.2 Data Storage Types

Local Storage

There are two primary types of local storage: RAM and ROM.

- **Random access memory (RAM)** can be read from and written to, by the applications. RAM is usually installed in every store-bought computer and it is volatile, or unstable, because it is used often to perform two actions (read and write) in the data processing stage of the IPO cycle.

  If a program crashes (ceases to work or encounters an error) or loses power while writing or saving the data, the information will be lost and not saved.

- **Read-only memory (ROM)** can be read only and cannot be written to at all. Due to read-only functionality, this memory is stable. The previously saved data is saved indefinitely, whether the device is powered on or off. Manufactured computers include prerecorded programs on ROM that cannot be changed and are used to power the device and perform complex basic functions (e.g., BIOS). Applications use ROM to read device settings and available memory options that the software can use.

External Storage

External storage can be used to supplement local storage or to archive information for long periods of time, freeing space from the local drive. There are many mediums, formats and structures for these external storage devices that vary greatly in storage capacities and cost.

Portable devices, such as a CD (compact disk) and DVD (digital video disc), are optical-media type read by laser and come in several variations: ROM to indicate read only, RW to indicate read and write, R to indicate one-time recording and RAM to indicate the ability to read, write and erase.

Universal serial bus (USB) flash drives are small, compact devices that can store a very large amount of data compared to their size. Named flash due to the data transfer speeds, the USB (i.e., USB stick, thumb drive) connects directly to a port that connects to the serial bus in the PC for quicker data transmissions. Like other technologies, USB generations are called USB types, e.g., A, B, C, 1.1, 2.0, and so on, and each one is faster than the other.

The challenge with portable data storage devices is that they can be easily lost and data can be compromised. Choosing a USB device with encryption can reduce the likelihood of some data breaches depending on the type of breach.

In most business applications, databases are commonly used to store complex data in an organized manner. Using text documents (i.e., flat files) with many types of data within each file can make it difficult to aggregate, query and access the text *within* the file (i.e., record) programmatically, because each file must be opened and read by a program, costing processing power and delaying results because flat files cannot be accessed concurrently.

In a database, each item is an attribute of the record and can be manipulated individually and collectively as part of a record. Additional database features include:

1. Powerful querying and reporting capabilities
2. Indexing for quicker access and search
3. Concurrent accessibility
4. Can update multiple records at once
5. Access control and greater security

For example, when a property manager adds a tenant to their application via the user interface, the application creates a record in the backend database with all the information the property manager entered via the user interface and creates a file copy of the lease for the tenant as a printout.

A relational database management system (RDBMS) is a relational model of a collection of information that is organized in tables and stored on a computer system. It is the most widely accepted database model today for saving information in an organized manner.

The overall purpose of a database is to store and retrieve related information. Database management systems (DBMSs) evolved from hierarchal, to network, to relational models. The relational database model has three major aspects—structures, operations and integrity rules.

Database information can be updated or modified via SQL queries to UPDATE or INSERT records one or many records at a time. The data structure attribution is in table rows and columns and describes data *in relation* to another piece of data in the database.

Although not all software applications use databases, a database is required for an organized, orderly access to meaningful data and is highly dependent on the data size and type being stored and the security level of the data. Finally, an Oracle database is a collection of data that is treated as a unit (e.g., property management system). Following are some key SQL statements with which one should be familiar, using the property management example.

2.4.3 Retrieving Information

A simple SELECT statement can be used as a wide net to get all columns from a single table:

SELECT *	**SELECT** is a SQL keyword that signifies read-only search * (asterisk) is a wild card to signify all information from that table
FROM Tenants;	FROM is a SQL keyword that signifies the location (in this case the Tenants table)

To retrieve specific records, a SELECT statement can be combined with JOIN, and WHERE statements where more than one table is used to obtain the records. There are four types of JOIN:

- **(INNER) JOIN**—Returns records that have matching values in both tables
- **LEFT (OUTER) JOIN**—Returns all records from the left table, and the matched records from the right table
- **RIGHT (OUTER) JOIN**—Returns all records from the right table, and the matched records from the left table
- **FULL (OUTER) JOIN**—Returns all records when there is a match in either left or right table

Example:

```
SELECT Leases.LeaseID, Leases.LeaseDate, Leases.Status, Tenants.TenantName

FROM Leases

INNER JOIN Tenants ON Leases.TenantID=Tenants.TenantID

WHERE Tenants.TenantName = 'John Smith' AND Leases.Status='A'
```

The Leases table is where all lease records are kept, with attributes such as lease dates, lease status (Active, Inactive), etc. The Leases table does not have a tenant's name—it only has a tenant ID—so a JOIN has to be made to find all leases that have a tenant name of 'John Smith' (names are only kept in the tenants table) and a lease status of active (kept in Leases table). Another way to search is to find John Smith's tenant ID and perform a search in the Leases table, only without a join.

Here is a modified **example:**

```
SELECT Leases.LeaseID, Leases.LeaseDate, Leases.Status

FROM Leases

WHERE TenantID = 6 AND Leases.Status = 'A';
```

Every field included in the SELECT portion of the query will show in the results. There is no need to include the field Leases.TenantID as every record returned will show the number '6'.

2.4.4 Updating Information

Updating records can be done much quicker via a SQL statement rather than one at a time via a GUI, although it can be extremely risky and the query should be vetted to ensure that the correct records are manipulated by performing several different SELECT statements first. Most often, the GUI includes several business rules that are checked for (e.g., not allowed to update name post record creation) that are nonexistent in most backend SQL statements. This applies to CREATE, UPDATE and DELETE statements.

Example: UPDATE Tenants

SET TenantName = 'Smithson'	All records matching the WHERE clause will be updated with Smithson.
WHERE LIKE "%smi%";	LIKE clause can be used with the SQL wildcard % to find all matching records, such as 'alsmi', 'smith', 'pasmian', etc.

% can be used in the front or back of every string to expand or reduce the search result.

2.4.5 Deleting Information

Deleting records should be done sparingly because an audit trail may not be captured and deleting may create orphaned records if proper cleanup is not performed to delete the joining records, because a relational database stores the information in multiple tables. For example, deleting a lease record will leave an active tenant with a nonexistent lease record, resulting in errors and possible GUI malfunctions. An orphan record is when a child record with a foreign key points to a parent record of a primary key field that no longer exists.

Example: DELETE FROM Tenants

WHERE TenantID=6;	Deleting a specific record with UID 6

Referential integrity is a property of data stating that all their references are valid. In the context of relational databases, it requires that if a value of one attribute (column) of a relation (table) references a value of another attribute (either in the same or a different relation), then the referenced value must exist.

Figure 2.11 provides an example of a database that has not enforced referential integrity.

Figure 2.11—Example of Lack of Database Referential Integrity

Tenants Table

TenantID	TenantName
6	John Smith
7	Louis Corola
8	Adam Williams

Leases Table

TenantID	LeaseID	LeaseStart	LeaseEnd	NumOccu
5	21	2018-07-26	2019-07-26	2
8	23	2020-01-30	2020-03-31	1
7	24	2020-02-10	2020-08-10	4
6	25	2020-04-22	2020-04-22	3

Database normalization is the process of structuring a relational database to reduce data redundancy and improve data integrity. Normalization entails organizing the columns (attributes) and tables (relations) of a database to ensure that their dependencies are properly enforced by database integrity constraints.

An example of data not normalized is depicted in **figure 2.12**.

Figure 2.12—Example of Nonnormalized Data

Tenants Table

<table>
<tr><th>TenantID</th><th>TenantName</th><th colspan="4">Leases</th></tr>
<tr><td rowspan="4">6</td><td rowspan="4">John Smith</td><th>LeaseID</th><th>LeaseStart</th><th>LeaseEnd</th><th>Status</th></tr>
<tr><td>25</td><td>2019-04-22</td><td>2019-10-22</td><td>I</td></tr>
<tr><td>31</td><td>2019-10-23</td><td>2019-12-31</td><td>I</td></tr>
<tr><td>36</td><td>2020-01-01</td><td>2020-12-31</td><td>A</td></tr>
<tr><td rowspan="2">7</td><td rowspan="2">Louis Corola</td><th>LeaseID</th><th>LeaseStart</th><th>LeaseEnd</th><th>Status</th></tr>
<tr><td>38</td><td>2020-06-01</td><td>2020-12-31</td><td>A</td></tr>
<tr><td rowspan="3">8</td><td rowspan="3">Adam Williams</td><th>LeaseID</th><th>LeaseStart</th><th>LeaseEnd</th><th>Status</th></tr>
<tr><td>33</td><td>2019-11-23</td><td>2020-05-31</td><td>I</td></tr>
<tr><td>35</td><td>2020-06-01</td><td>2020-12-31</td><td>A</td></tr>
</table>

A more normalized equivalent of the structure above might look like **figure 2.13.**

Figure 2.13—Example of Normalized Data

Tenants Table

TenantID	TenantName
6	John Smith
7	Louis Corola
8	Adam Williams

Lease Table (query is ordered by TenantID)

TenantID	LeaseID	LeaseStart	LeaseEnd	Status	NumOccu
8	23	2020-01-30	2020-03-31	I	1
8	33	2019-11-23	2020-05-31	I	1
8	35	2020-06-01	2020-12-31	A	1
7	24	2020-02-10	2020-08-10	I	4
7	38	2020-06-01	0202-12-31	A	4
6	25	2020-04-22	2020-04-22	I	3
6	31	2019-10-23	2019-12-31	I	3
6	36	2020-01-01	2020-12-31	A	3

The most commonly used databases are described in the following paragraphs.

2.4.6 Common Databases

The most commonly used databases are described in the following paragraphs.

Oracle

Oracle, the name synonymous with RDBMS, tends to set the bar for other database management tools. Oracle, a very large company with decades of experience, offers database management tools that are incredibly robust with numerous features. Oracle database resource overhead requires powerful servers or cloud version and the willingness to keep up with consistent hardware upgrades. Some small and mid-range businesses opt out due to the high cost and are willing to forgo the rich feature set for a more comparable lower-cost or free database version.

Microsoft SQL

Similarly, Microsoft SQL is a mature product that is both fast and stable, and it offers performance management tools and overhead resource adjustments, allowing the database administrator to optimize resources (e.g., network activity and memory). Moreover, Microsoft SQL is complementary to the full Microsoft ecosystem of products allowing for seamless integrations.

Disadvantages to consider are that the cost is prohibitive to most small and midrange organizations and, even with performance tuning, Microsoft SQL Server requires more resources when compared with other databases.

MySQL

One of the most alluring features of mySQL is it is free to use and offers a competitive feature set, one being multiple user interfaces from which to choose. It offers integrations with the costlier databases, such as IBM DB2 and Oracle. As with any free version, it requires more manual work, because it lacks some features that are basic in the other databases and features that other systems perform automatically, like create incremental backups. Additionally, there is no built-in support for XML or online analytical processing (OLAP), and support would involve a cost.

PostgreSQL

This is another free database management option that is used mostly for web application databases. PostgreSQL supports JSON (JavaScript Object Notation) and is scalable to handle terabytes of data. It does offer numerous interfaces and built in functions to get started quickly. However, as with any free tool, it relies heavily on forums and sparse, or outdated, documentation to self-serve if support is needed, which is likely during the configuration phase, which can be confusing to some.

MongoDB

MongoDB is also referred to as a NoSQL database that is fast and easy to use. It supports JSON and other document-based storage, which may be in a multitude of data structures, such as semistructured and unstructured data. It does, however, require learning a new proprietary query language, because SQL is not used, and tools are available to enable the translation from SQL to MongoDB query language. Moreover, configuration is critical because if set incorrectly or left at default, it can pose a security risk.

IBM DB2

IBM DB2 is at the same level as Oracle and Microsoft SQL in cost structure, which leaves out most smaller businesses. Moreover, at an additional cost, third-party tools and software are required to make clusters or multiple secondary nodes function. DB2 offers cloud and onsite setups, and multiple jobs can be automated using Task Scheduler, freeing the database administrator to perform other tasks.

Page intentionally left blank

2.5 Chapter 2 Knowledge Check

REVIEW QUESTIONS

1. Which of the following is an example of an input?

   A. Monitor
   B. API
   C. Printer
   D. Speakers

2. Choose the **BEST** example of a string data type.

   A. String
   B. A
   C. 30,100
   D. Hello World

3. You want to test for a condition in your program and you do not know how many times the code will need to test for it—which is the **BEST** control flow to use in this case?

   A. While loop
   B. If else
   C. For loop
   D. Do While loop

4. The extension ".ini" is an example of which of the following?

   A. RAM
   B. Background activity
   C. Flat file storage
   D. ROM

5. Which of the following is an example of a reserved key word used for control flows in programming?

   A. If
   B. SELECT
   C. WHERE
   D. f(x)

Answers on page 90

Chapter 2 ANSWER KEY

Review Questions

1. A. Monitor
   B. API. Refer to page 63.
   C. Printer
   D. Speakers

2. A. String
   B. A
   C. 30,100
   D. Hello World. Refer to page 64.

3. **A. While loop. Refer to page 72.**
   B. If else
   C. For Loop
   D. Do While loop

4. A. RAM
   B. Background activity
   C. Flat file storage. Refer to page 80.
   D. ROM

5. **A. If. Refer to page 69.**
   B. SELECT
   C. WHERE
   D. f(x)

SOFTWARE DEVELOPMENT FUNDAMENTALS
An ISACA Certificate

Chapter 3:

Software Development Life Cycle

Software Development Life Cycle

Software Development Life Cycle

3.1 Learning Objectives

After completing this chapter, learners will be able to:

1. Recognize ISO/IEC/IEEE standards (e.g., 12207:2017, 15289:2011, 730-2014, 90003).
2. Describe SDLC management terms and concepts, including roles and responsibilities.
3. Identify the five software development phases.
4. Describe software management methodologies (waterfall and Agile).
5. Recognize Agile development methodologies (Scrum, XP, FDD, DevOps).
6. Illustrate how to choose a software methodology.
7. Explain software versioning and release best practices.
8. Describe configuration management.
9. Describe code repositories and proper code branching.
10. Explain DevSecOps and Secure SDLC.
11. Identify project management tools and techniques.
12. Express software development ethics.

3.2 Overview

The International Organization for Standardization (ISO), International Electrotechnical Commission (IEC) and the Institute of Electrical and Electronics Engineers (IEEE) created and refined a long-standing guidance for initiating, planning, controlling and executing many interdependent processes in software engineering. Each of the following standards focuses on a part of the software development field and holistically integrates the disciplines to create common terminology, processes, techniques and practices that are discussed in this section and are appropriate regardless of management methodology.

- Systems and software engineering-software life cycle process of ISO/IEC/IEEE 12207:2017
- Information content requirements of ISO/IEC/IEEE 15289:2011
- Software quality assurance processes of IEEE Std 730-2014
- Software engineering—Guidelines for the application of ISO 9001:2015 of ISO/IEC/IEEE 90003, which involves quality management standards and guidelines for software development

3.2.1 ISO/IEC/IEEE 12207:2017

This standard establishes a set of processes for managing the life cycle of software. The life cycle of software releases includes the following versions:

- **Alpha**—The first release of the software package by the development team that is usually to be used by the test team. Alpha version may or may not include all the features, but it includes the suggested architecture, design and basic functionality to identify any major design flaws.
- **Beta**—Likely includes all features and requirements said to be part of the application and includes major bug fixes from the prior Alpha version. The Beta version is also to be used by the test team for various testing methods and gates.

- **Release candidate (RC)**—The most stable version of the software and usually includes all features and bug fixes from preceding versions. This version is customarily delivered to testing and the user acceptance testing (UAT) phase specifically and, in some cases, to selected customers willing to test the prereleased version and provide input.
- **Release to market (RTM)**—The build version that is released to the general public and will often not have any priority 1 bugs.

The purpose of this standard is not to prescribe a specific software life cycle model, modelling approach, method, technique or development methodology but, rather, distinguishes between a *process* and *stage*.

- **Process**—interrelated or interactive activities that transform inputs into outputs. Processes occur within stages or phases.
- **Stage (i.e., phase)**—A defined period of time within a life cycle of an object that communicates the state or position of its realization within that life cycle. Stage-gates are points in time when a program is reviewed and a decision is made to commit expenditures to the next set of activities on a program or project, to stop the work altogether, or to put a hold on execution of further work.

ISO/IEC/IEEE 12207:2017 recognizes that software life cycles differ and may be divided into four main process groups: agreement, organizational project-enabling, technical management and technical processes.

3.2.2 ISO/IEC/IEEE 15289:2011

ISO/IEC/IEEE 15289:2011 is based on the life cycle processes specified in ISO/IEC 12207 and provides requirements for identifying and planning the specific information items (i.e., information products, documentation) to be developed and altered during systems and software life cycles and service processes.

It specifies the purpose and content of all identified systems and software data records and life cycle information items, as well as records and information items for information technology service management.

The information item contents are defined according to generic document types (description, plan, policy, procedure, report, request and specification) and the specific purpose of the document.[39]

3.2.3 IEEE Std 730-2014

This standard, harmonized with ISO/IEC/IEEE 15289:2011 and ISO/IEC/IEEE 12207:2017, is responsible for software quality assurance processes (SQA). Requirements for initiating, planning, controlling and executing the SQA processes of a software development or maintenance project are established in this standard.

Figure 3.1 displays how the Software Quality Assurance process fits in to SDLC and Project Management processes illustrating the multi-layer approach needed. It depicts the relationship between the organization and the software project planning, activities, tasks and the closing of the project. Moreover, it shows the context of the SQA activities within the formal project lifecycle of initiating, planning, monitoring, controlling, and closing. Software quality assurance activities are executed within the planning, monitoring, and controlling part of the project lifecycle, whether it is for acquiring off the shelf software or developing it in house.

[39] International Organization for Standardization (ISO), *ISO/IEC/IEEE 15289:2015 Systems and software engineering – Content of life-cycle information items (documentation)*, www.iso.org/standard/66422.html

Figure 3.1—Software Quality Assurance and Project Planning

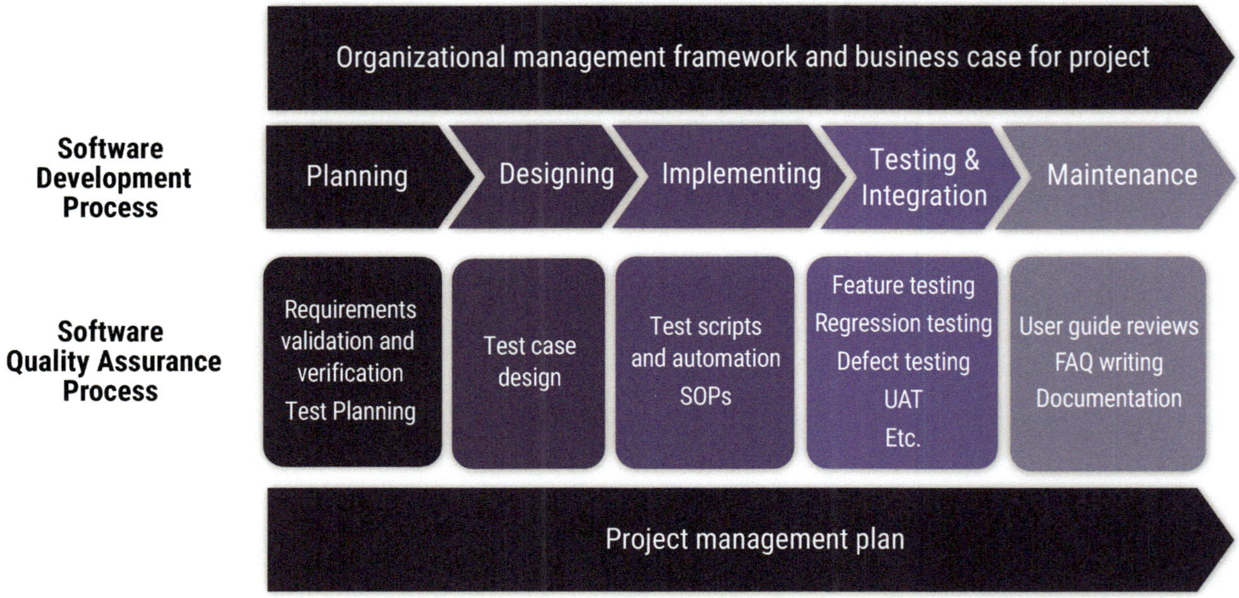

3.2.4 ISO/IEC/IEEE 90003

This international standard employs the process approach, which incorporates the plan-do-check-act (PDCA) cycle and risk-based thinking. The PDCA cycle (**figure 3.2**) enables an organization to ensure that its processes are adequately resourced and managed, and that opportunities for improvement are determined and acted on.[40]

The software engineering guidelines for the application of ISO 9001:2015 to computer software can be briefly described as follows:

- **Plan**—Establish the objectives of the system and its processes, and the resources needed to deliver results in accordance with customers' requirements and organization policies, and identify and address risk and opportunities.
- **Do**—Implement what was planned.
- **Check**—Monitor and (where applicable) measure processes and the resulting products and services against policies, objectives, requirements and planned activities, and report the results.
- **Act**—Take actions to improve performance, as necessary.[41]

[40] International Organization for Standardization (ISO), ISO 9001:2015(en) *Quality management systems – Requirements*, www.iso.org/obp/ui/es/#iso:std:iso:9001:ed-5:v1:en

[41] International Organization for Standardization (ISO), ISO/IEC/IEEE 90003:2018(en) *Software engineering – Guidelines for the application of ISO 9001:2015 to computer software*, www.iso.org/obp/ui/#iso:std:74348:en

Figure 3.2—Plan-Do-Check-Act

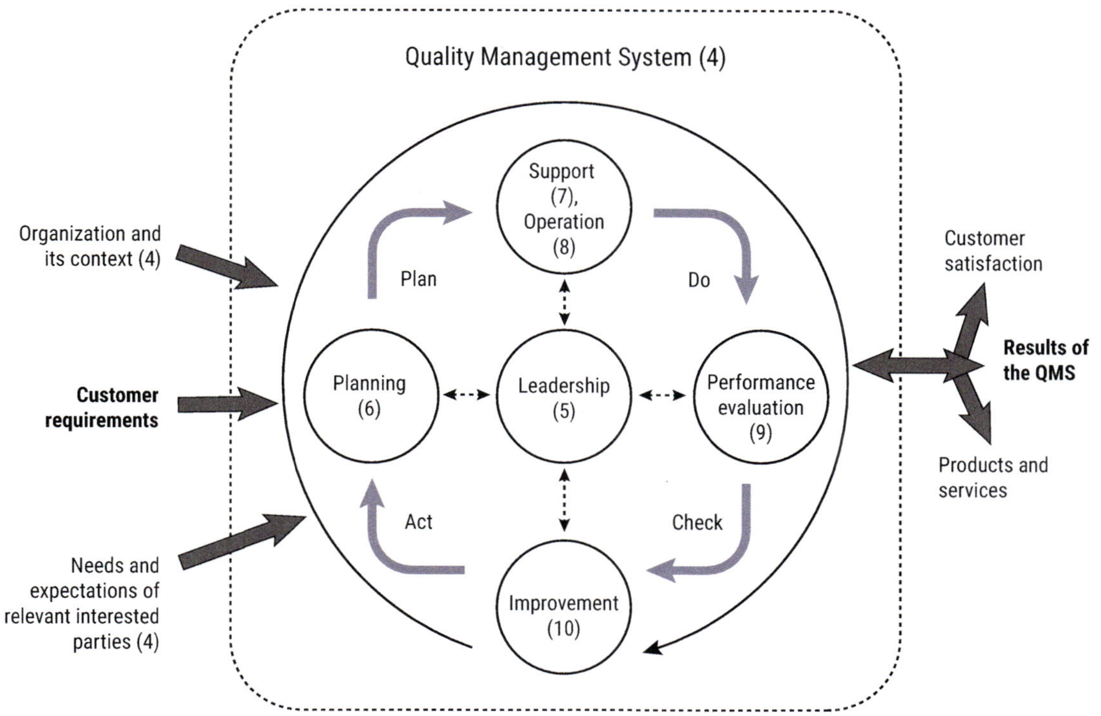

Source: ISO, "ISO/IEC/IEEE 90003:2018(en), Software engineering —Guidelines for the application of ISO 9001:2015 to computer software", figure 2, www.iso.org/obp/ui/#iso:std:74348:en. ©ISO. This material is reproduced from ISO 9001:2015 with permission of the American National Standards Institute (ANSI) on behalf of ISO. All rights reserved.

The software development lifecycle process (SDLC) is the period of time that begins with the decision to develop a software product and ends when the software is delivered. This cycle typically includes a planning phase, design phase, implementation phase, test phase and maintenance phase (**figure 3.3**).

Figure 3.3—Software Development Life Cycle (SDLC)

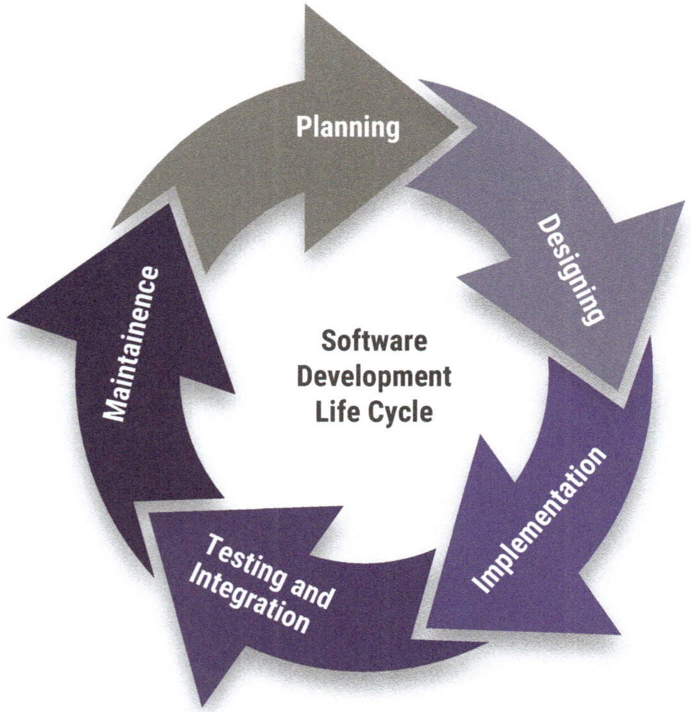

Software selection and acquisition is not a phase in the standard SDLC. However, if a decision was reached to acquire rather than develop software, software acquisition is the process that should occur after the planning and requirements gathering phase.

In the case of a software acquisition, the implementation phase, where the development and coding are performed, consists mostly of configuration and setup of the acquired software.[42]

3.3 SDLC Phases

3.3.1 Planning Phase

The planning phase, also referred to as the analysis, initiation or requirements gathering phase, includes the following high-level activities:

- **Gathering and analysis of requirements**—All requirements must be clear and allow for an unambiguous solution.
- **Scheduling and team assembly**—Schedules and resource plans are created using different tools and templates depending on the methodologies selected. In iterative methodologies, schedules may range from one-to-six weeks in length, while traditional methodologies may take much longer. Resources are assigned roles although titles and tasks differ from one methodology to the next. Two commonly used management methodologies are detailed in the SDLC Management section (section 3.4).
- **Create communications plan**—How the stakeholders will communicate, and *what* interval and modalities are used are detailed to set expectations.

[42] ISACA, *CISA Review Manual, 27th Edition*, USA, 2019, www.isaca.org/bookstore/cisa-exam-resources/crm27ed

- **Risk identification**—Identify risk and its potential impact on the business, as well as mitigation techniques.
- **Benchmarking**—Identifies capabilities when contrasted with industry standards and best practices or internal comparisons with other similar projects.

3.3.2 Software Requirements Analysis

The input, processing, output (IPO) paradigm, as introduced in the previous chapter, is used in every phase of the software development life cycle.

The requirements phase has its own inputs, processing mechanisms and outputs, which are later used as inputs to the design phase (**figure 3.4**). Fundamentally, a requirement (i.e., acceptance criteria) is criteria that a system or its components must satisfy to be accepted by the end user or customer.

Figure 3.4—Overview of IPO Process of Software Requirements

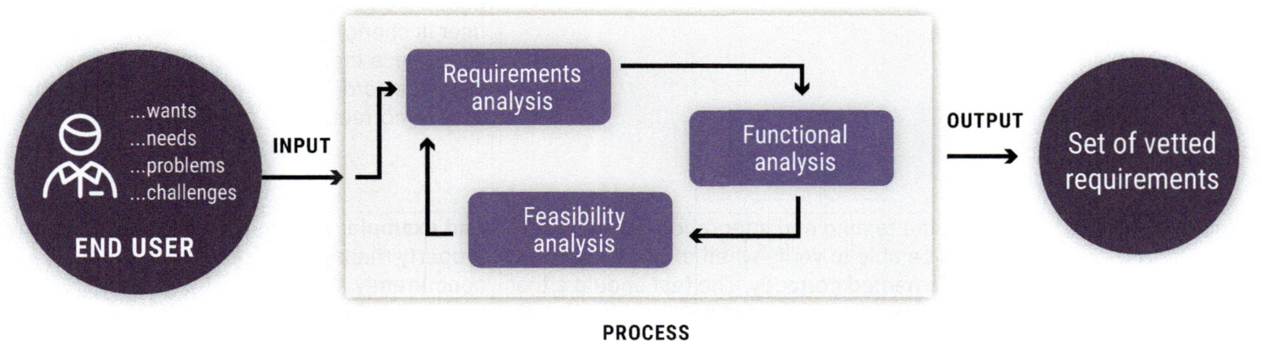

Characteristics of a Proper Requirement

Following are the characteristics of proper requirements (**figure 3.5**):

- Clear (concise, simple, defined) and understandable
- Unambiguous
- Testable (verifiable)
- Correct
- Necessary and nonredundant
- Feasible (realistic, possible)
- Independent and complete
- Atomic

Figure 3.5—Requirements Characteristics		
Requirement Characteristic	**Explanation**	**Examples**
Clear (concise, simple, defined) and understandable	Requirements should not contain unnecessary verbiage or information.	**Bad example:** Sometimes, the property manager will enter city name, which the system will understand, but, sometimes, the closest city may replace it if it is a similar spelling, so the user does not need to know what the zip code is, and it will still be understood by the system. **Good example:** As a property manager using the *search* capability, I should be able to enter a zip code or city name, and the application will return (show results) of all managed properties based on either a zip code or a city name.
Unambiguous	There should be only one way to interpret the requirement. All acronyms should be spelled out.	**Bad example:** The system shall not accept passwords shorter than eight characters. **Good example:** The system shall not accept passwords shorter than eight characters when a user is changing his/her password. If the user enters less than eight characters while choosing the password, an error message will ask the user to correct it.
Testable (verifiable)	During the testing and integration phase, testers should be able to verify whether the requirement is implemented correctly. The test should either pass or fail. It is good practice to not leave any term or sentence up for interpretation.	**Bad example:** The system should allow many property managers to perform searches concurrently. **Good example:** The system should allow a minimum of 10 concurrent users performing searches at once (users=both property managers and tenants).
Correct	If a requirement contains static facts, these facts should be applicable in all related use cases; otherwise, each should be a separate requirement.	**Bad example:** Lease renewals and monthly rent will show all applicable taxes (including 6% state tax). **Good example:** Lease renewals and monthly rent will show all applicable taxes according to the state/zip code of the property location (tax depends on the state).
Necessary and nonredundant	A requirement is not necessary if it is not needed by any stakeholders or if removing the requirement will not cause loss of functionality. Moreover, the same feature request (e.g., a pop-up calendar) will not be created multiple times in the requirements document or user stories. It is better to list the location of the feature being tested as details in the one requirement.	For example, a developer may think that adding a visual, interactive map of all the properties may be a nice feature (which it is), but it is not needed nor asked for. All requirements should be traced back to a customer or business request.
Feasible (realistic, possible)	The requirement should be doable within existing constraints, such as budget, schedule and available resources.	For example: The application should support every available currency—this is not feasible nor needed if the properties are located in North America, so US dollars should be the only supported currency because testing, schedules and resources would need to be expanded to support other, less likely, currencies for this use case.

Figure 3.5—Requirements Characteristics *(cont.)*

Requirement Characteristic	Explanation	Examples
Independent and complete	To understand the requirement, there should not be a need to know any other requirement.	**Bad example:** It should be sorted by price. **Good example:** Search results screen: all unoccupied apartments search by zip code or city should be sorted by price (low-high).
Atomic	When in doubt break down the requirement into multiple one feature element. This is better organization and traceability for each of the phases in the software development lifecycle; especially testing and maintenance.	**Bad example:** The system will provide the tenant options to sign the lease, pay rent, submit a repair request, and provide information about local restaurants. **Good example:** req1. The system will provide the tenant options to sign the lease electronically. req2. The system will provide the tenant options to pay current rent due via visa credit card. And so on...

When writing requirements, each one should be traced back/linked (also known as requirements traceability) to a customer or organizational request. More importantly, and in addition to the characteristics outlined here, the written requirement should accompany a screen design or any additional artifacts or links to related requirements to create context. Lastly, communication and collaboration with the team is key, and written requirements do not replace that interaction.

Search capability is a standard requirement for most end-user facing applications. However, that is too broad; for example: Search where? What can I search on?

Figure 3.6 depicts one way of how to order and link a few of the search requirements, using the property management application as an example.

Figure 3.6—Requirements Traceability

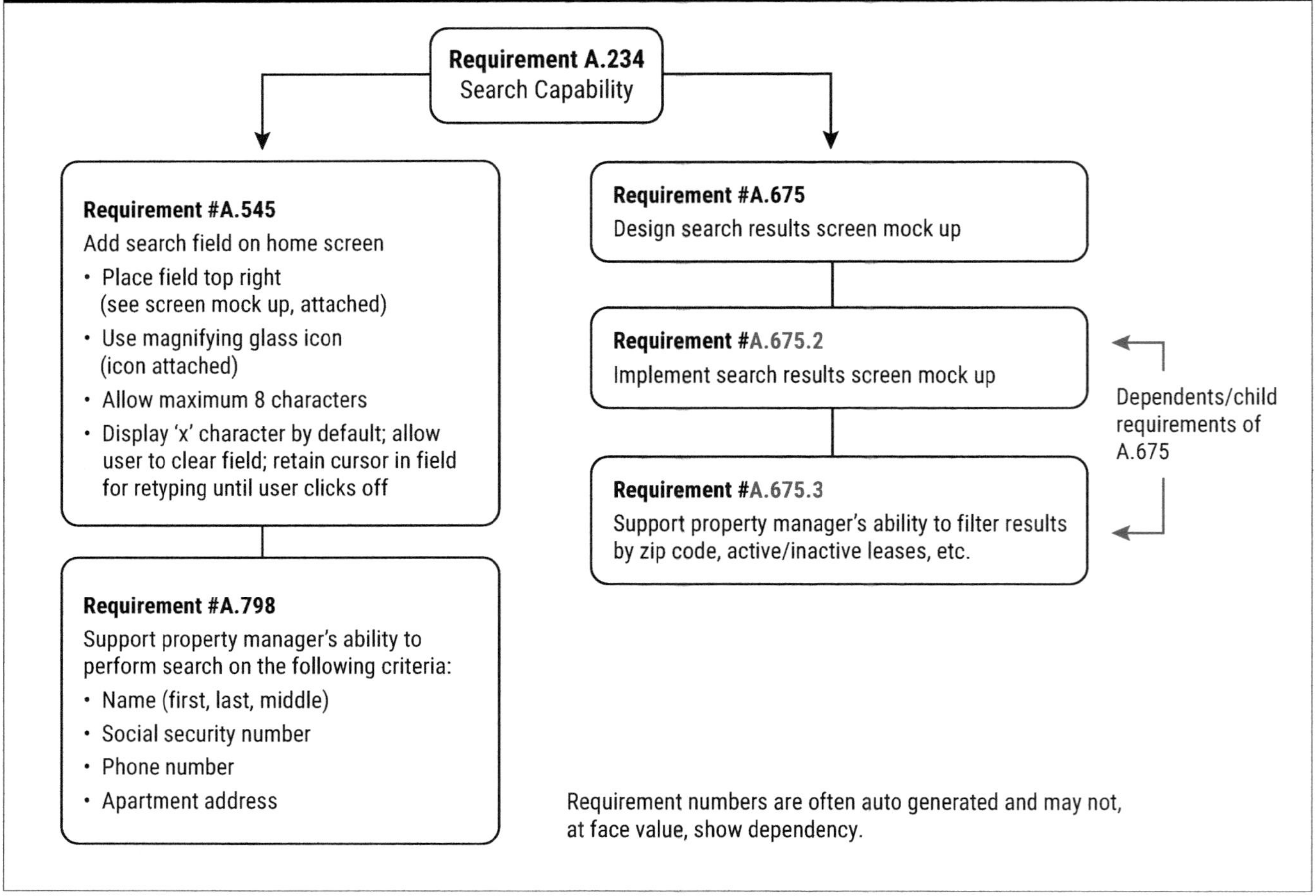

3.3.3 Scheduling and Team Assembly

Considering the enterprise context in the industry and the needs of the proposed software application, commitment by leadership to establish the support structure needed for the project execution is important. This includes allocating resources with the proper competencies, defining a RACI chart (RACI charts outline who is responsible, accountable, consulted and informed for a task), outlining decisions, setting priorities, agreeing on funding and approving value objectives.

3.3.4 Roles and Responsibilities

Software development is primarily conducted by coders, software developers, programmers and software engineers. These roles interact and overlap, and the dynamics between them vary greatly across development departments and enterprises. The SDLC, however, includes many of the following roles that also change titles (not necessarily responsibilities) based on the methodology used:

- Software engineers, programmers, developers, or coders:
  - Write source code to program computers for specific tasks, using languages, such as Java, Python and C++, to carry them out.
  - Apply engineering principles to build software and systems to solve problems. They use modeling languages and other tools to merge databases, create websites, write scripts, design algorithms and much more.

- Software architects:
  - Responsible for facilitating architectural design and implementation decisions of a program—the blueprint of the program is extremely important because it dictates fundamental structures, such as security, scalability and integrative flexibility evolution. Moreover, they assist the developers in adhering to and designing their code to comply with the overall architectural vision.
- Product owners, business analysts (PO/BA):
  - Responsible for soliciting, collating, refining and translating business requirements to development idioms. This role is also sometimes called subject matter expert (SME) and can assist with prioritization.
  - They are a great resource to review an alpha version and help with organizing and testing during user acceptance testing (UAT) before it reaches the end user.
- Project manager, scrum master (PM/SM):
  - Project manager, in the traditional role, is responsible for managing all aspects of a project and has full functional autonomy over the project resources in a matrixed environment.
  - They are also responsible for project charter, schedules and budget. Contrast that with a scrum master, who has no autonomy and leads by negotiating and coaching the team to embrace Agile values and principles, and encouraging the team to become self-organizing.
- Test engineers, testers and quality analysts (QA/QC):
  - This role is pivotal for any software development project. Testers truly become experts in the resulting software, because they are well acquainted with how it works by attempting to break it.
  - Testers are responsible for defect and bug creation, and, in some cases, documentation to the end user in the form of a user manual.
  - There are two different types of analysts: quality assurance (QA) is concerned mainly with processes which affect quality, while quality control (QC) is concerned with testing the product for quality.
- Database Administrators (DBA):
  - This role is responsible for database selection, architectural design, queries, scripts, database integration, security and any reporting needs.

3.3.5 Communication Plan

Regardless of which software development methodology is used, a communication plan is needed. The communication plan is designed to inform stakeholders and end users of the project throughout the project life cycle, with varying degrees of information, dependent on the users consuming the information. A communication plan may include several or all of the following methods:

- **Kickoff meeting**—Usually at the project commencement; this meeting is used to mark the beginning of the project. Representative stakeholders are usually in attendance and are informed of approved scope, schedules, requirements, purpose, resources and, in some cases, the allocated budget for the project.
- **Information repository sharing**—It is best practice to create a central repository to house all test plans, templates, requirements documentation and many other project artifacts in a central document repository that provides version control, organization and, most importantly, access control that is dependent on user access level.
- **Defect prioritization meetings**—These meetings are used to address defects (bugs), assign a level of severity and priority to each of the defects, then a decision may be made to allocate them to an upcoming sprint depending on the priority.
- **Change control**—Any requirement changes are documented and tracked.
- ***Ad-hoc* meetings**—These meetings are used when a specific topic needs to be addressed quickly. They are not scheduled far in advanced nor recurring.

The communication plan should have cadence and mediums defined up front to set both team and stakeholder expectations and ensure project transparency throughout the software development life cycle. This will prevent assumptions, and create an atmosphere of transparency and accountability. Some enterprises opt for many face-to-face meetings repressing the project and inducing bottlenecks that can arise from scheduling conflicts (more time in meetings, less time for actual work) instead of creating a self-serving, independent team with an always available repository of reports, current status and all details pertaining to the project.

3.3.6 Risk Management

A risk management plan is part of every project planning phase, and a software application development project is no different; the risk register is a listing of items that can impact the project adversely. Once identified, the risk is then analyzed, categorized, monitored and controlled.

To produce the risk list, a brainstorming session begins by asking questions, such as: What kind of issues will the project encounter? Are there any regulations we should be concerned with? Do we have the resources with the skills needed for this project? The following examples include other common areas that should be taken into consideration when creating a risk register:

- Technical
- Cost
- Schedule
- Financial
- Environmental
- Client
- Contractual
- Regulation
- Political
- People

When considering risk, the *likelihood* of it occurring should be considered, based on historical and experiential basis and that requires gathering historical data to gauge properly. The impact of the given risk, however, should dictate the level of analysis and data collection compared to time spent.

Risk categorization is two-dimensional: impact (severity if this event occurs) and likelihood (probability of event occurring). The rating scale for each is shown in **figure 3.7**:

Figure 3.7—Risk Categorization

Impact (Consequences)	Likelihood (Probability)
5-High/Severe: causes some aspect of project failure	**5-High:** expected to occur
4-Medium/High: threatens project	**4-Medium/High:** likely to occur
3-Medium: requires some aspect to adjust (scope, budget, resources)	**3-Medium:** could occur
2-Medium/Low: threatens secondary aspect of project	**2-Medium/Low:** unlikely to occur
1-Low: very low consequence	**1-Low:** exceptional, rare possibility

Risk can be:

- Mitigated (e.g., a solution is implemented)
- Transferred (e.g., insurance is purchased or a third-party vendor is used instead)
- Avoided (e.g., the architecture may be replaced by a less risky one)
- Accepted

Risk should be documented in a risk register (**figure 3.8**).

Figure 3.8—Sample Risk Register

ID	Date Raised	Risk Description	Probability	Impact	Owner	Action (Mitigated, Transferred, Avoided, Accepted)	Status
1	03/28/2020	Onsite servers purchase, management and maintenance require resources to manage.	High	High	Adam	**Transferred**—Decided to purchase cloud storage for database and software hosting instead of onsite strategy.	Closed
2	03/28/2020	New software application will be used by end users that have never used a computer before.	High	Medium/ High	Mary	**Mitigated**—A new training and change management group has been assembled to create a robust training program.	In progress
3	03/28/2020	Third-party payment processing vendor contract negotiations are taking longer than expected.	High	Medium	Trent	**Avoided**—May need to incorporate our own payment gateway instead of using third party—further investigation is needed.	In progress
4	04/15/2020	Scope creep of 17 new requirements from legal that must be incorporated to accommodate new contractual and regulatory mandates.	High	High	Julie		New

3.3.7 Benchmarking

Benchmarking is a standard or a baseline against which measurements and comparisons can be made. It defines a reference point on which all progress will be measured to identify either advancement or a regression. Benchmarking is especially helpful when deciding whether to create the desired software from the ground up or opting to purchase a software application and integrate it with existing applications.

Best-of-breed (BoB) decision-making exercise can help make that decision by comparing costs and maintenance, risk involved with integrations, and one-to-one requirement capabilities analysis.

Benchmarking can be broken down into several practical steps:

- Compare software proposal to existing software out in the market, in the same industry sector. Compare costs, maintenance and warranty, upgrade options, and accessibility and contrast that with in-house-built software and the overhead required to plan, design, implement, support, document and maintain it.
- Perform a business flow analysis by observing the current manual process that is performed by the end users. Determine if there is a way to solve or automate a large percentage of the tasks with existing enterprise software. If yes, a simple script or two may be able to support the business automation requested for the software, or adding several features to an already created software can minimize costs and reduce maintenance.

3.3.8 Designing Phase

The designing phase is the period of time in the software life cycle during which the designs for architecture, software components, interfaces and data are created, documented and verified to satisfy requirements.

This phase is crucial, because it translates documented requirements (the *what* phase) to a physical representation that can be interacted with, however limited (the *how* phase). It can be accomplished in the following several ways, depending on need.

- **Architectural diagrams**—These diagrams can include numerous architectures at different layers of the system that include the application. A diagram may include security layers (see **figure 3.9** for example), communication and network layer, software design layer, database, software integration layers, and more.

Figure 3.9—Example Security Architectural Diagram

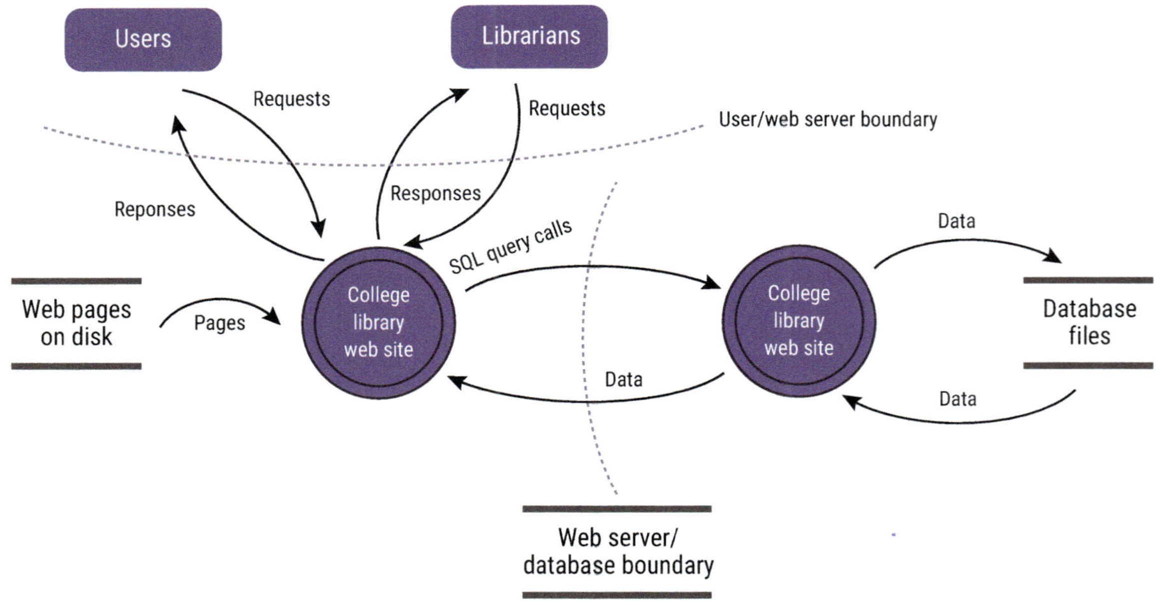

Source: OWASP. The figure is available under the Creative Commons Attribution ShareAlike 4.0 International license.

- **Entity relationship diagrams (ERD) for databases**—This diagram depicts a set of real-world entities and the logical relationships among them. An ERD can be used to display the structure of the tables that will store the data elements and attributes, and it also provides a visual starting point for the database design that is needed for the application and the database (**figure 3.10**). The database is used by, and for, the application.

Figure 3.10—Sample Entity Relationship Diagram

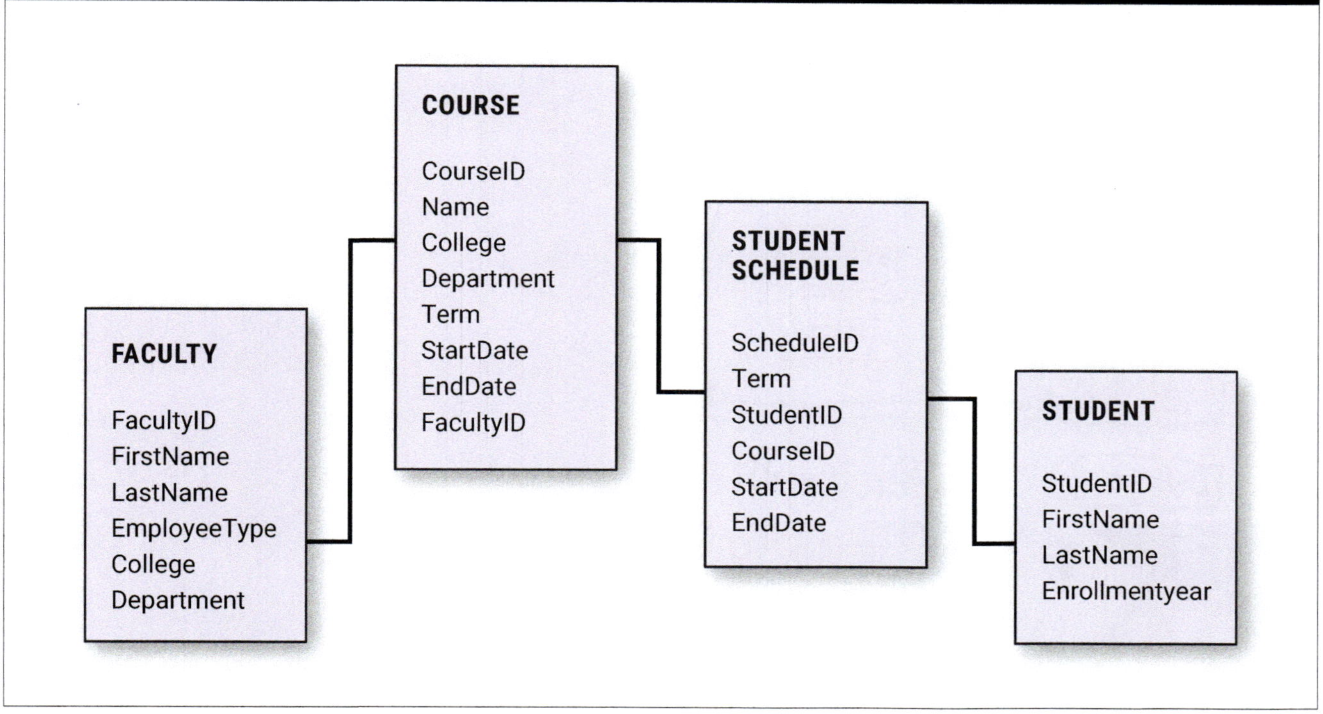

- **Prototyping**—This design method is likely the most cost effective. Prototyping is the process of rapidly assembling a working application model to test various aspects of a design, illustrate ideas or features, and gather early user feedback. Prototyping uses programmed simulation techniques to represent a model of the final system to the user for guidance and critique. The emphasis is on end-user screens and reports. Internal controls and full working backends are not priority items, because this is only a model.

  One example, paper prototyping, uses paper compared to actual coding, which is slower and more expensive each time a redesign is needed for the intended user interface. After the screens of the application are drawn (can be done electronically now, as well), they are placed in functional order and the end user is observed as they interact with the paper designs.

  The key difference between mere sketches and a paper prototype is interactivity. A paper prototype is brought to life by a design team member who simulates what the computer would do in response to the user's clicks and keystrokes, by rearranging pieces, writing custom responses and occasionally announcing some effects verbally that are too hard to show on paper.

  Because a paper prototype is actually interactive, one can essentially user-test it: give users a task to do and watch how they do it.[43] For example: If the observer notices that the end user looks for the Submit button for longer than a few seconds, then the design will likely be improved (**figure 3.11**).

  There are four design dimensions to consider:

  - Fidelity indicates how similar the design is to the finished interface and is one dimensional. Low-fidelity prototypes omit details, use cheaper materials or use different interaction techniques. High-fidelity prototypes are very similar to the finished product.
  - Breadth refers to the fraction of the feature set represented by the prototype. A prototype that is low-fidelity in breadth might be missing many features, having only enough to accomplish specific tasks.
  - Look and feel

[43] MIT EECS, "Reading 8: Prototyping," http://web.mit.edu/6.813/www/sp17/classes/08-prototyping/

Figure 3.11—Sample Prototype Designs

Source: Glassman, E.; P. Guo; D. Jackson; D. Karger; J. Kim; R. Miller; S. Mueller; C. Sims; H. Zhang; "Reading 8: Prototyping," MIT EECS, http://web.mit.edu/6.813/www/sp17/classes/08-prototyping/. The figure is available under the Creative Commons Attribution ShareAlike 4.0 International license.

Skeleton programs and prototypes are more time intensive and require an actual program to be coded at a very high level with superficial features. A prototype will likely not have all features, but rather the design concept in raw working form. For example, a prototype may also be used to show how an integration works or that the software can query another system.

3.3.9 Implementation Phase

The implementation phase, also called the execution phase, is where the development and coding are done. The requirements, coupled with the design framework, are translated to source code. All the necessary functions, algorithms, data structures, etc. are implemented in this phase. The development and coding are performed in a preproduction or development environments, sometimes referred to as sandboxes, or testbeds.

If a waterfall methodology is used, this phase can range from a month to years (depending on scope of project and size of team) because all the requirements are coded. However, if an iterative approach is used, then a subset of the requirements will be implemented within each sprint (a sprint is a timebox schedule of one month or less during which the team produces a potentially shippable product increment) so the customer can gain access to a portion of the requirements instead of having to wait for the full set of requirements to be implemented all at once.

Figure 3.12 depicts the differences between an iterative and sequential model and the repetitiveness of the phases within each.

Figure 3.12—Waterfall vs. Agile Workflows

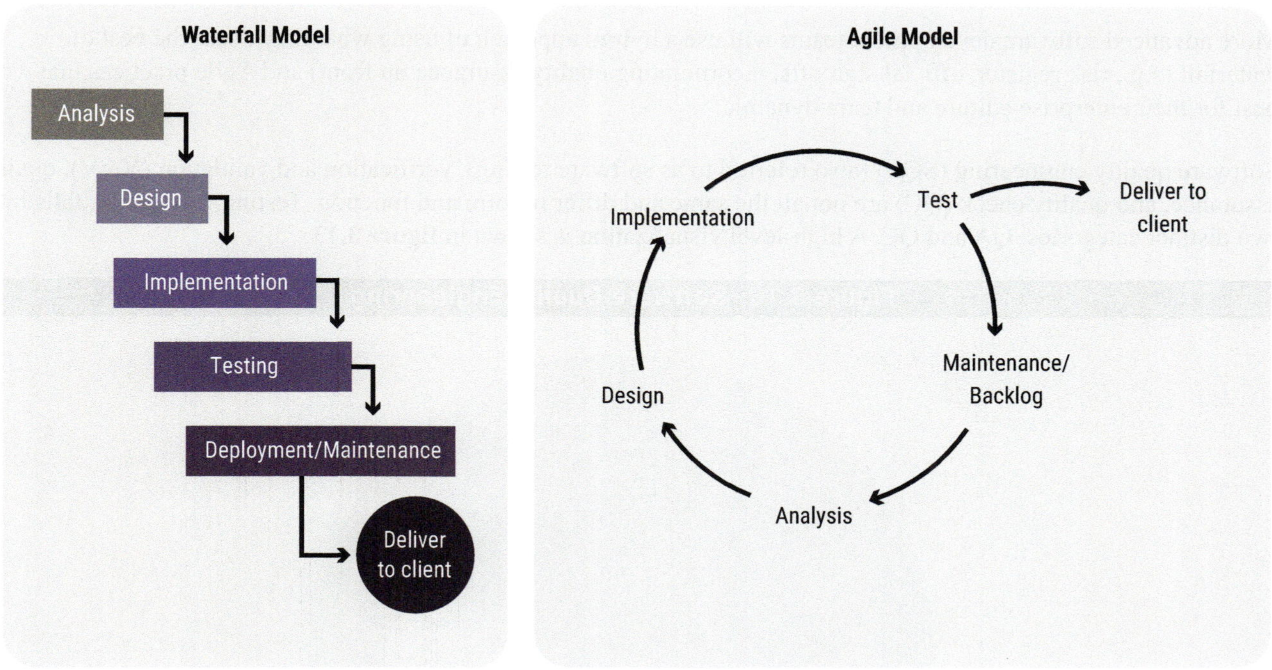

Source: Adapted from Trivedi, P.; A. Sharma; "A comparative study between iterative waterfall and incremental software development life cycle model for optimizing the resources using computer simulation," 2013, 2nd International Conference on Information Management in the Knowledge Economy, Chandigarh, 2013, pp. 188-194

Additional tasks performed during the implementation phase are:

- **Unit tests creation**—The purpose of the test is to ensure that the internal operation of the program performs according to specification. It uses a set of test cases that focus on the control structure of the procedural design of the code. Developers normally write their own test cases at the software or hardware function (low level) level.
- **Code commenting**—Best practice is to notate the code as it is being written to ensure accuracy and completeness.

3.3.10 Testing and Integration Phase

During the fourth phase of the software development life cycle, the application is tested thoroughly to identify and uncover any noncompliance requirements, unhandled errors or functional scenarios (i.e., use-cases that have and have not been considered during the planning phase), and defects (i.e., bugs).

Testing is usually performed in a staging environment. A staging environment (also called testbed) is a server location with simulated data and a release candidate copy of the application. After testing has occurred, these environments can be rolled back (and duplicated) to pretesting condition for the next phase of testing, because data may get corrupted while testing new versions of the application.

The two main SDLC methodologies—Agile and waterfall—may perform the same test types, but testing is executed at different times in the SDLC. In Agile, testers are part of each sprint and work concurrently with software development at each phase; whereas, in waterfall, testing is formally performed after the implementation phase. Waterfall is costlier due to this siloed approach because testers' input can offer an advantage by writing test cases during the planning phase, hence covering scenarios that have not been previously considered.

More advanced software development teams will use a hybrid approach of using what they deem the best of waterfall (e.g., risk register, official sign offs, incorporating quality assurance up front) and Agile practices that work best for their enterprise culture and team dynamic.

Software quality engineering (SQE) (also referred to as software testing), verification and validation (V&V), quality assurance, and quality check (QC) are not all the same and differ in form and function. Testing, in general, falls into two distinct categories: QA and QC. A high-level visualization is shown in **figure 3.13**.

Figure 3.13—Software Quality Engineering

Source: ASQ, "Differences Between QA and QC," https://asq.org/quality-resources/quality-assurance-vs-control

3.3.11 Quality Assurance Responsibilities

Quality assurance is responsible for establishing a quality management system (QMS) and adherence to ISO 9001:2015, the international standard specifying requirements for quality management systems. It is the most prominent approach to quality management systems across innumerable number of industries—not just software development.

According to the American Society for Quality (ASQ), a QMS is defined as a formalized system that documents processes, procedures and responsibilities for achieving quality policies and objectives. A QMS helps coordinate and direct enterprise activities to meet customer and regulatory requirements and improve its effectiveness and efficiency on a continuous basis.[44]

Quality assurance is process-oriented and is involved starting with the planning phase, throughout the SDLC, and postproduction of the systems. They primarily focus on eliminating process variation by designing and creating guidelines, analyzing past results, and revising a set of tightly and precisely defined processes and procedures. Furthermore, ensuring these guidelines are *followed exactly* and are strictly implemented leads to ensuring the final quality of the product.

QA is accountable to both the business management and the end user, and oversees the SDLC process as all inputs and outputs for each of the phases can affect the end result. Several techniques are employed to identify and reduce possible defects in the released software, as described in the following paragraphs.

- Create detailed documentation, such as standard operating procedures (SOPs), checklists and process flow charts. Determine grading hierarchy and perform periodic performance audits to check for conformance.
- Investigate defects, issues, etc. that are found to deviate from standards
- Improve development practices by implementing code reviews, test automation, and early planning and coordination with development and testing during earlier phases
- Defect root cause analysis and prevention

Many teams also incorporate Kaizen in every aspect of the business workflow and software management. Kaizen is a Japanese term coined by Toyota, the car manufacturer, which means Kai-change and Zen-good to refer to continuous improvement. The goals of continuous improvement include the elimination of waste, defined as activities that add cost, but do not add value; just-in-time (JIT) delivery; production load leveling of amounts and types; standardized work; paced moving lines; and right-sized equipment.

A closer definition of the Japanese usage of Kaizen is "to take it apart and put it back together in a better way." What is taken apart is usually a process, system, product or service.[45]

Kaizen is a daily activity whose purpose goes beyond improvement. It is also a process that, when done correctly, humanizes the workplace, eliminates hard work (both mental and physical), and teaches people how to do rapid experiments using the scientific method and how to learn to see and eliminate waste in business processes.

Kaizen emphasizes making small changes by forming small feedback loops to understand what needs to change—constantly driving towards the goal of highest quality.

The 5 Whys is an interrogation technique of asking, "Why?" at least five times, to reach the root cause of an issue. It can be used as a proactive measure to test requirements for their value, and as a reactive approach to reach the fault. **Figure 3.14** shows an example using the property management application.

[44] American Society for Quality (ASQ), *Quality Glossary*, https://asq.org/quality-resources/quality-glossary/q
[45] Hafey, R.; *Lean Safety: Transforming Your Safety Culture with Lean Management*, Kaizen Blitz Team Charter

Figure 3.14—Example of 5 Whys Interrogation Technique

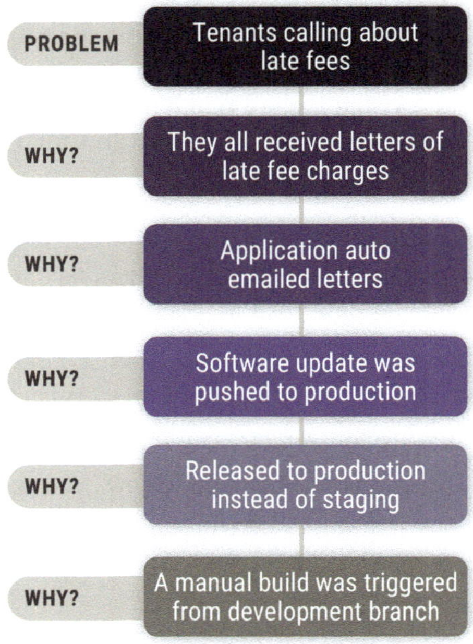

Root cause analysis is a reactive technique that is used to identify the true cause of a failure. Whether it is a service failure or a defect in an application, the fundamental causes should be identified and corrected to prevent recurrences. The Ishikawa diagram was created by Kaoru Ishikawa, an engineering quality management professor. The Ishikawa diagram, also known as Fishbone Diagram, is a visual representation for problem analysis (**figure 3.15**).

When a problem arises, it may be easily attributed to human error when it could have been caused by a combination of inputs, such as faulty controls, ambiguous procedures and processes, technology limitations, or environmental complexities. Knowledge of the faulty entry points and pin-pointed correction can ensure that the problem does not repeat.

Figure 3.15—Sample Ishikawa Diagram

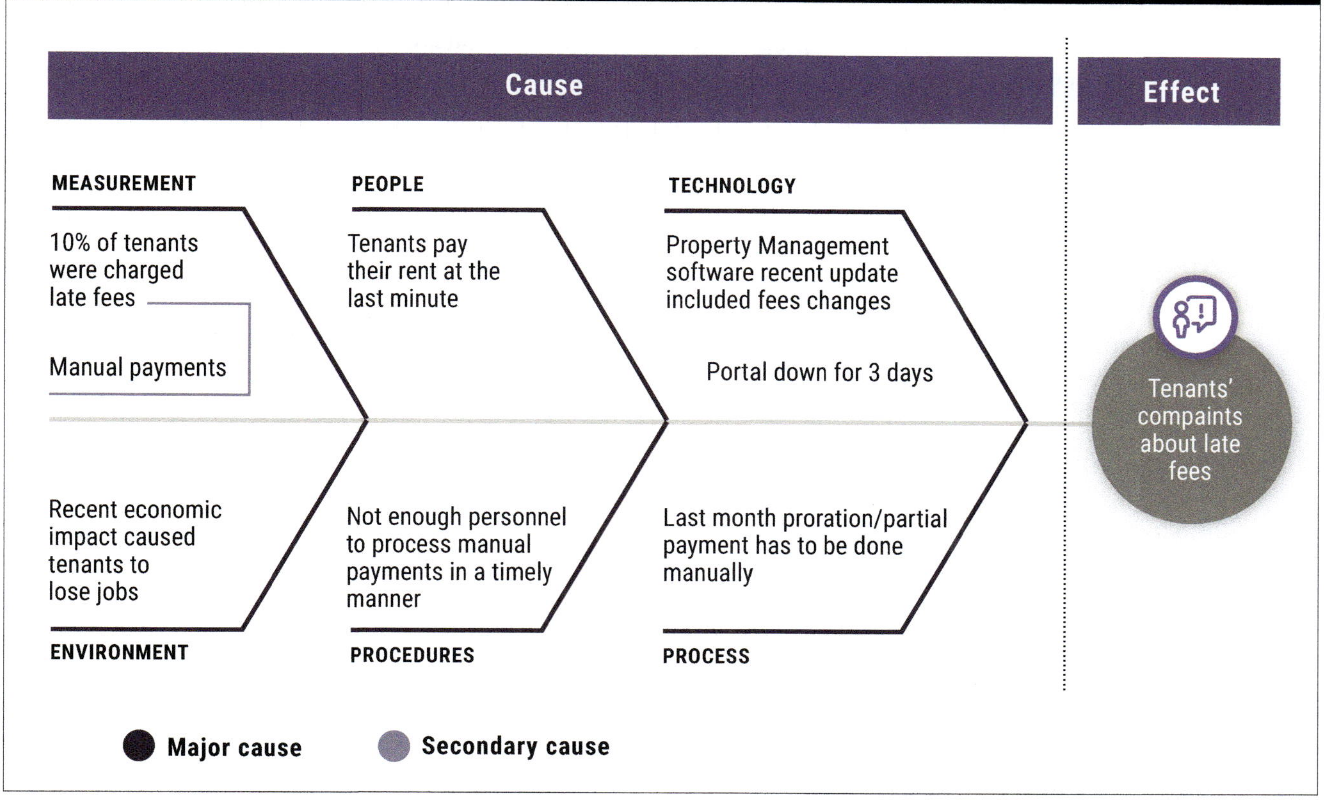

3.3.12 Quality Check Responsibilities

Quality check testing is the process of analyzing a software item to detect the differences between existing and required conditions (i.e., gap analysis, bugs) and to evaluate the features of the software items.[46] QC is responsible for thoroughly testing the system and software application to ensure it complies with the functional requirements, performance, operability, compatibility and regulatory or security requirements. Any defects (i.e., bugs) found are entered into a defect tracking system and prioritized before being assigned a developer for correction.

The tests performed cover the several dimensions of the software as described, performance, functionality, compliance, etc., but the test specifics are concerned with the following for each of the dimensions mentioned:

- **Boundary value testing**—Every data value that corresponds to a minimum or maximum input is tested to ensure the code instruction sets and decision trees allot for no outside boundary values. A boundary value is a value that lies at, or just inside or just outside, a specified range of valid input and output values. Choices often include maximum, minimum and parameters such as *equal to or*.

  For example:

  Input acceptable values of integers 0 to 30 should not allow any integers <0 or >30.

  Performance of minimum 5 sec of response time should not be >5.

- **Branch testing**—Each time a decision is made in code via the flow of controls structures, such as *if, then*, a test is created and executed to ensure that, for each decision point, each possible code branch is executed at least

[46] IEEE, *IEEE 729-1983 IEEE – Standard Glossary of Software Engineering Terminology*, 18 February 1983, https://standards.ieee.org/standard/729-1983.html

once. Moreover, this branch coverage produces enough test cases so that each decision has a true and a false outcome at least once.

- **Critical control point**—This test type is concerned with function, or an area in a process or procedure, the failure of which, or loss of control over, may have an adverse effect on the quality of the finished product and may result in an unacceptable performance.

  For example:

  A total of 15 users should be able to query the database concurrently. A test example will be testing with an incremental increase of users to identify the breaking point where the application performance can be gauged. At 16 users, the application becomes very slow to respond, at 18 users it becomes non-responsive, etc.

- **Coverage analysis**—Testers write and execute test cases by invoking program structural elements to determine the adequacy of an instruction set against quality benchmarks. Coverage analysis is needed when attempting to execute each statement, branch, path or iterative structure in a program.

  They ensure coverage of the full source code by employing several methodologies:
  - **Black-box testing**—This technique of testing involves testers who do not have any knowledge of the interior workings of the application nor the code. The tester is oblivious to the system architecture and will interact with the system user interface by providing inputs and examining outputs, without knowing how and where the inputs are worked on or which code branches are stimulated to provide the outputs. This approach is used by all end users and consumers when using any released application and by the stakeholders when user acceptance testing (UAT) is performed during the SDLC development process.
  - **White-box testing**—Also called glass testing or open-box testing, is the detailed investigation of internal logic and structure of the code. Testers are likely doing code walk throughs, architectural design reviews and other analysis to look inside the source code and find out which unit/chunk of the code is behaving inappropriately (also known as debugging).
  - **Gray-box testing**—Also known as translucent testing, is when the tester has limited knowledge of the source code inner workings. The limited known information may be flow diagrams, at minimum, and can be performed by testers and end users alike.

3.3.13 Testing and Integration Automation

Manual Testing

Most software testing cycles are lengthy and exhaustive due to manual execution of the test cases. If the project schedule falls behind, the test cycle is the first to be shortened. The tester mimics the role of an end user and uses the software as a customer would use it, to identify any unexpected behavior or defects.

There are multiple stages for manual testing that correlate to the SDLC phases. Each of these testing types incorporates aspects of performance, security, usability and integration aspects of the application, as described throughout this guide (e.g., security in DevSecOps or Secure SDLC sections) that the team may choose to automate or continue manual execution:

- **Unit testing**—This is usually performed by developers during the design and implementation phases.
- **Integration testing**—During the implementation and testing and integration phases, all the parts of the applications are tested together (database setup, network conditions, application) via integration testing or regression testing.
- **Regression testing**—Also referred to when all modules affected by a fixed defect are tested to ensure no other logic or area of the program is adversely affected.

- **System testing**—Also known as smoke testing, system testing is an initial pretest to ensure that the platform and application are ready and stable for rigorous testing cycles. System testing is usually performed numerous times during the testing and integration phase.
- **Exploratory testing**—Sometimes referred to as gorilla testing or *ad hoc* testing, exploratory testing is an undocumented, unplanned testing where anything goes. If an unexpected or odd behavior is encountered, a defect is logged for further discussion. This is the most creative type of testing and many new feature ideas usually arise from this approach.
- **User acceptance testing (UAT)**—Tested by the stakeholders, end users and SMEs, UAT is performed at the end of the testing and integration phase. This testing is more concerned with the day-to-day business flow requirements only.

Automation Testing

Automation testing is also known as test automation and is usually performed by testers with development experience and skills at writing test automation code—writing code-to-test-code. The tester usually uses another software testing application in which to write the scripts and execute the test cases during off hours via an automated job.

Test automation is especially helpful for automating regression and system test cases because they are redundant, boring and continuous, resulting in higher costs and time spent.

Test automation can greatly reduce time and cost; however, it comes with its own challenges and defects. The cost savings and accuracy that can be gained ultimately add value to the project. The overhead (e.g., setup time and maintenance) required to upkeep the test scripts can be time consuming; if the application changes an architectural aspect, all test scripts may need to follow suit.

3.3.14 Defect Analysis

The two main parameters that form the basis for effective defect tracking are defect severity and defect priority. Severity describes the gravity of an undesirable occurrence. Hence, when it comes to defects, the severity of a defect indicates the effect it has on the system in terms of its impact, especially on the quality of the software application.

Priority indicates the comparison of two or more items or conditions, where one has to be given more importance than the other(s) and has to be tackled/resolved first, before proceeding to the next one(s). Therefore, in the context of defects, the priority of a defect indicates the urgency with which it needs to be fixed.

Defect severity is assigned by the test analyst, and then defects are usually reviewed by the product owner or team to determine prioritization. Defect severity is determined based on the impact (and probability) the defect has on the quality of the software application.

The software defect assessment comprises several dimensions that ultimately affect its quality:

- Impacted user experience
- Occurrence (interval)
- Performance benchmark (e.g., stress testing)
- User Interface (e.g., misspellings and truncation)
- Regulatory (e.g., Americans with Disability Act (ADA), ISO 9001, and HIPPA compliance)
- Security (e.g., user access control and safeguards of data)
- Integration testing (e.g., multiple application communication, and regression)

Severity is normally categorized in one of four levels:

- **S1 Critical**—A defect that completely hampers or blocks testing of the product/ feature is a critical defect. This includes if the application crashes, or it becomes unusable or not able to proceed further, and/or any catastrophic system failures leading the user to not being able to use the application.
- **S2 Major**—Any major implemented feature that is not meeting its requirements/use case(s) and behaves different from expected can be classified under Major severity. This includes when the functionality grossly deviates from the expectations or not doing what it should be doing.
- **S3 Moderate**—A defect that still exhibits some deviation from certain criteria or does not meet requirements specification. However, functionality as a whole is not affected.
- **S4 Low**—A minor low-severity defect occurs when there is almost no impact on the functionality, but it is still a valid defect that should be corrected.

After test analysts set the severity and initial priority, the team or the product owner reviews the bug details, reproduction steps, attached log files, screenshots or videos displaying the error and sets a priority.

Typically, there are four priorities from which to choose:

- **P1 Critical**—Generally, all priority 1 bugs must be fixed prior to the software release. Some examples of a priority 1 issue are regulation or security noncompliance, omitted high priority requirements, incorrect calculations, and system or application crashes.
- **P2 High**—High-priority items include not-found errors, and innocuous messages that are misleading.
- **P3 Medium**—The defect is specific to a certain environment, user or browser, or it includes very specific (sometimes convoluted) steps to reproduce the issue.
- **P4 Low**—Examples include cosmetic issues and minor misspellings.

Dependent on the schedule and budget, medium and low priority defects are sometimes mitigated with documentation in some cases.

The priority and severity settings of application defects are not only dependent on the application users, but also on the industry in which the software will be used. Some industries are more regulated than others (medical, financial, etc.) and noncompliant functionality must be fixed prior to releases, to avoid fines and legal battles.

Software testers can become true experts of the software application, probably more than any other role on the team. They begin by testing as a customer would, attempting to perform the functional requirements as outlined, not knowing what or how the code is written.

Test plans and test cases are created to cover every scenario for each requirement, resulting in hundreds of test cases depending on the complexity. Continuing with the earlier example of a property management application, see **figure 3.16** for example test cases.

Figure 3.16—Sample Test Cases

REQ1: The system will not accept passwords less than eight characters. If the user enters less than eight characters, an error message will ask the user to correct it.

Test Case #	Input	Expected Output	Actual Result	Test Results	Notes
1	12345	Message requesting correction should appear	Password accepted; no message	FAIL	<8 characters
2	12345678 9101232	Message requesting correction should appear	Message appeared as: no correct.	FAIL	Need product owner to provide acceptable customer facing message.
3	12345678	Password should be accepted; no message	No message appeared. Password accepted.	PASS	=8 numeric characters accepted
4		Message requesting correction should appear	No message appeared. Password accepted.	FAIL	NULL password Empty passwords are a security concern.
5	768GHTFD	Password should be accepted; no message	No message appeared. Password accepted.	PASS	=8 numeric characters Alphanumeric is accepted.
6	7!(*&89A	Password should be accepted; no message	Application crashed; unhandled exception.	FAIL	=8 characters Special characters handling missing.

The last phase of software testing involves the customer, or their representatives, via user acceptance testing (UAT). The main intention of UAT is to validate the end-to-end business flow. It does *not* focus on any other test type that was previously performed and considered complete. User acceptance testing is carried out in a separate testing environment with a production-like data setup usually called *staging*.

3.3.15 Deployment

The release of the software is performed after the development process is completed and is most often considered to be part of the development process rather than deployment process. It includes all the steps needed to prepare any system for the installation of the software and, in some cases (especially windows applications), may include prerequisite software that may need to be installed—one such example is the .NET framework. A good application includes all needed prerequisites as part of the installation.

When the new application is replacing an existing application, four strategies for transitioning to the new system can be used:

- **Parallel operation**—Both the new and old systems are operating for a given period. While this approach is likely the most expensive and time-consuming because two systems have to be supported and maintained for a set time, it is the safest, since both are accessible and output can be produced and compared.
- **Direct changeover (direct cutover)**—The old system is shut down and the new one takes over processing at once. This is the least expensive and least time-consuming transition, but also the riskiest, because the new system cannot be reverted to the original.
- **Pilot conversion**—One branch, department or division at a time is converted to the new system. Experience gained from each installation is used to the benefit of the next one. One disadvantage is the prolonged rollout time.
- **Phased rollout**—One function of the new system at a time is placed into operation. The advantage of this strategy is allowing users to learn one part of the system at a time.

When the back-end database is not to be reused with the new application, a data migration might be needed. Data migration is a very important consideration because the data may need to be manipulated to fit the new database model used by the new application. Data migration should be carefully planned and tested, and appropriate methodologies and tools should be utilized to minimize the risk of:

- Disruption of routine operations
- Violation of the security and confidentiality of data
- Conflicts and contention between legacy and migrated operations
- Data inconsistencies and loss of data integrity during the migration process.

Because not all new application deployments go as planned, good practice dictates that the tools and applications required to reverse the deployment are available prior to attempting the production deployment.

3.3.16 Maintenance Phase

After testing is completed, it is common to have numerous defects that have been deprioritized; however, a successful software release is software that meets the customer's requirements with no priority-1 bugs.

When the software is released to production, additional releases will likely follow with all fixes for priority-2 bugs at minimum. The defect write-up is used as the requirement because it includes the use-case-specific steps to recreate the issue. After all fixes are checked in and a release candidate is created, the testing can again commence.

Maintaining the system is an important aspect of SDLC because it requires long-term commitment. Some software applications have been around for decades with continuous bug fixes and enhancements prolonging the longevity and stability of the application.

In most instances it is the responsibility of the support or help desk team to provide specialist knowledge of the production systems to identify and assist in system change/development and problem resolution. Support is mostly triggered when a help desk call is initiated by a user and then escalated based on the complexity of the issue and the level of expertise required to resolve the problem. Some help desk/support calls can result in a change request being submitted to correct the problem/defect. A change request can also be used to request additional functionality in the system. Change requests are handled through the change management process.

3.3.17 End of Life

In some cases, the software architecture becomes outdated and the time to sunset an application come due. Sunsetting, retirement phase, or end of life (EOL) are terms often used when the application is no longer supported by its creator.

One example of this is the Windows operating system. Windows 7 has been active for 11 years and was retired during January 2020. The millions of PCs using Windows 7 are now faced with a choice to either upgrade to a new operating system requiring a learning curve, or continue using Windows 7 past the EOL date. Patches and fixes will no longer be released and the operating system will become increasingly more vulnerable to security issues.

3.4 SDLC Management

As with any project initiation to solve a problem, it is wise to have a methodology; a set of phases that delineate the work in a common-sense way that helps the team arrive at the solution in an organized manner.

According to the Project Management Institute (PMI), methodology is defined as a system of practices, techniques, procedures and rules used by those who work in a discipline.[47]

Researchers surveyed project managers and other team members about their choice of methodologies. The results indicate that although methodologies such as Agile Scrum are more prevalent than 10 years ago, traditional

[47] Project Management Institute (PMI), *A Guide to the Project Management Body of Knowledge: PMBOK® Guide, Sixth Edition,* 2017, www.pmi.org/pmbok-guide-standards/foundational/pmbok

methodologies, including the waterfall model, are still popular. Enterprises are also taking a hybrid approach, using multiple methodologies on projects. Furthermore, their choice of methodologies is associated with certain organizational, project and team characteristics.[48]

There are many methodologies to choose from; this guide focuses on the two most well-known methodologies in software development—waterfall and Agile. All methodologies include some form of planning, designing, testing and maintenance.

3.4.1 SDLC Models—Waterfall

Waterfall methodology is the oldest and, therefore, considered the traditional software development process. It was introduced by Dr. Winston W. Royce in 1970.[49] It is named waterfall due to its linear, sequential execution much like a waterfall's water descends in a top-down manner.

This model of the software development process includes constituent activities. Typically, a concept phase, requirements phase, design phase, implementation phase, test phase, and installation and checkout phase, are performed, in that order, possibly with overlap, but with little or no iteration.[50]

Waterfall is a defined process control model that encompasses a stage-gate, procedure-focused development cycle with formal sign-off at the completion of each level or phase. Waterfall affords enough time for each phase to ensure every piece of information is fully vetted, known and understood.

This methodology ensures repeatability and predictability before moving on to the next phase (**figure 3.17**). Any identified deviation is considered a defect in waterfall. This is in contrast with Agile, which is a lean development methodology that is a nondeterministic approach—pieces of information are understood and expected to not be fully known (expect the unexpected)—where deviations are exploited as possible features and opportunities for innovation.

Figure 3.17—Sequential Processes Illustration

Waterfall is considered by some to be a rigid model because the steps have to be executed in the exact order they are prescribed, and the activities are usually siloed for each role. Following are the phase descriptions:

- **Requirements**—Requirements gathering is done first and in great detail. Moreover, the manner in which the requirements are written use specific keywords, such as:
  - **Must**—The specification is an absolute, e.g., application must be ADA compliant
  - **Must not**—For example, once clicked, response time must not exceed 5 seconds
  - **Should**—This specification is a recommendation. There may exist valid reasons to ignore this requirement, but the full implications must be understood and carefully weighed before choosing a different course, e.g., application should be less than 45 MB.

48 Vijayasarathy, L.R.; C.W. Butler; "Choice of Software Development Methodologies: Do Organizational, Project, and Team Characteristics Matter?," *IEEE Software*, vol. 33, no. 5, pp. 86-94, Sept.-Oct. 2016

49 Royce, W.; "Managing the Development of Large Software Systems," https://web.archive.org/web/20170513073313/http:/www.cs.umd.edu/class/spring2003/cmsc838p/Process/waterfall.pdf

50 *Op cit* IEEE, *IEEE 729-1983 IEEE – Standard Glossary of Software Engineering Terminology*

- **Should not**
- **May**
- **May not**

After the requirements are gathered and documented, they are frozen and no changes can be made, to enter the design and implementation phase.

- **Design**—The requirements are an input to the software architecture design (models, business rules and schema).
- **Implementation**—Actual development and coding.
- **Testing**—Software testing, both quality assurance and quality check for defect discovery and fixing (bugs are sent to development to work on fixes only).
- **Deployment and maintenance**—Installation and deployment to the field with continuous maintenance and bug fixes.

When using waterfall, one can expect very long project times with an abundance of documentation for each of the steps; standard operating procedures (SOPs), test-template formalities, and approvals are common occurrences.

Many government agencies used to, and some still do, use this methodology. Its application today is more of a hybrid approach, combined with Agile, allowing for more flexibility with a reduction in the documentation burden it once required.

3.4.2 SDLC Models—Agile

Agile software development has dominated the latter half of the past 50 years of software engineering and is now the mainstream software development method of choice worldwide.[51] The ability to be flexible, fast to respond, and collaborative are common words used to describe Agile.

Agile software development is an umbrella term for a set of frameworks and practices based on the values and principles expressed in the *Manifesto for Agile Software Development* and the 12 principles behind it.[52]

As shown in **figure 3.18**, the following 12 principals summarize the Agile values.

[51] Hoda, R.; N. Salleh; J. Grundy; "The Rise and Evolution of Agile Software Development," *IEEE Software*, vol. 35, no. 05, pp. 58-63, 2018
[52] Agile Alliance, "What is Agile?," www.agilealliance.org/agile101/

Figure 3.18—Agile Values

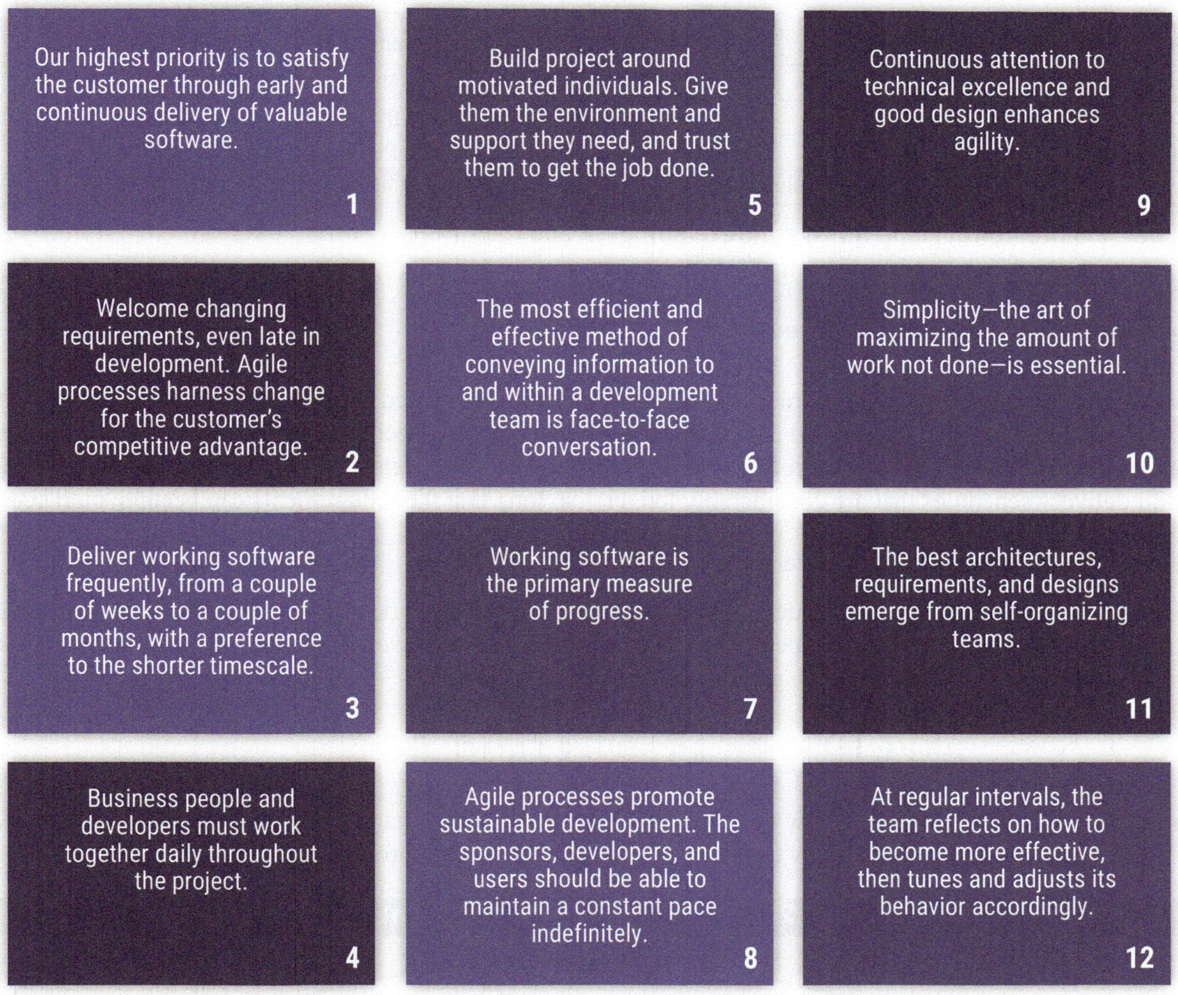

Agile emphasis is on *how* people work—the value is in the end result and the expediency to respond quickly to changing market needs, while creating a culture of enterprise values where change is a constant—not an anomaly.

3.4.3 The Agile Manifesto

The Agile Manifesto depicted in **figure 3.19** further reinforces the Agile commitment to its values.[53]

[53] "Manifesto for Agile Software Development," 2001, https://agilemanifesto.org

Figure 3.19—Agile Manifesto	
We are uncovering better ways of developing software by doing it and helping others do it. Through this work we have come to value:	
Individuals and interactions	Over processes and tools
Working software	Over comprehensive documentation
Customer collaboration	Over contract negotiation
Responding to change	Over following a plan
That is, while there is value in the items on the right, we value the items on the left more.	
Source: "Manifesto for Agile Software Development," 2001, https://agilemanifesto.org	

3.4.4 Agile Development Frameworks

Software delivery schedules have become shorter; there is an increased need for faster development cycles and more visibility and collaboration with the end user. All this is translating into specific Agile practices that have proven to shorten the feedback loop from the customer and provide software that meets the customers' needs now, unlike waterfall delivery schedules that are considerably longer—so long that the requirements can become outdated.

Under the Agile umbrella, there are several frameworks:

- Scrum is a framework for product development that helps the team work together. The term originated in rugby known for cross functioning, high performing teams that *huddle*, i.e., work together to win the game, unlike waterfall, which can be likened to a relay race.
- Extreme programming (XP)
- Feature-driven development (FDD)

Figure 3.20 shows a common workflow of a development sprint. Notice that, compared with waterfall, this cycle includes *all* phases *within* a sprint every time—not once and done for the project.

Figure 3.20—Sample Development Sprint Workflow

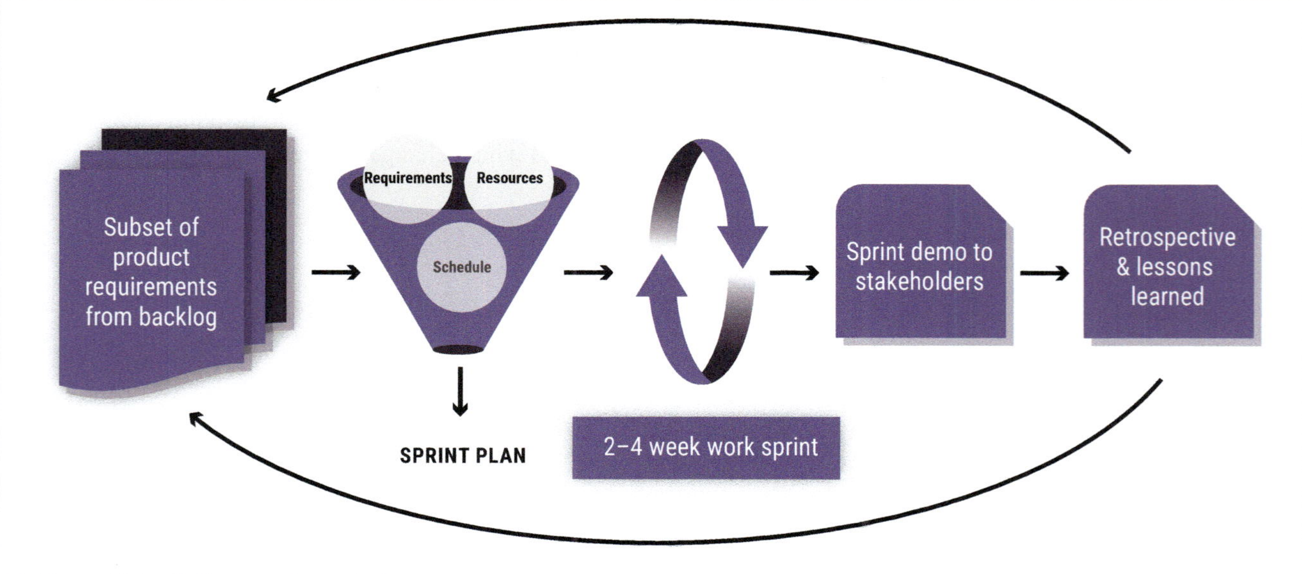

Moreover, the Agile consortium of frameworks offers many techniques that can be leveraged across all three depending on what the team prefers, project needs and the enterprise culture. Agile-specific terms are detailed in the following subsections.

3.4.5 User Stories

In waterfall, the customer's needs are called requirements; in Agile, they are referred to as user stories. User stories differ greatly from requirements because they incorporate an actor—who or what will benefit from this requirement being implemented? It forces the author of the requirement to think of the value that this user story must provide in terms of usability and function (**figure 3.21**). The actor may be a system process or an end user, but the actor's role helps ensure the user story can be tested (i.e., testable, actionable value-add). Following are the components and the questions that should be answered, from the user's (customer's) perspective, when writing a user story:

- As a (***who*** wants to accomplish something?)
- I want to (***what*** does the user want to accomplish?)
- So that (***why*** does the user want to accomplish that thing?)

Figure 3.21—Example Technical and Human Actor Responses

Human Actor Role	Technical Actor Role
• **As a** resident renter • **I want to** be able to pay my rent online • **So that** I do not have to be concerned with office hours and personnel availability.	• **As a** payment processing function • **I want to** be able to verify, accept and process online payments via Visa, Mastercard and American express only. • **So that** payments are processed and properly reflect accurate balances on the customer's and business accounts.

3.4.6 Assigning Effort Points

Traditional software teams provide estimates in a time format: days, weeks and months. Many Agile teams, however, have transitioned to story points measurements. Story points rate the *relative effort* of work in a **Fibonacci sequence** format: 0, 1, 1, 2, 3, 5, 8, 13, 21, 34, 55, etc. Fibonacci sequence is the addition of two preceding numbers (e.g., 21+34=55).

It may sound counter-intuitive, but that abstraction is helpful because it pushes the team to make tougher decisions around the difficulty of work, without getting bogged down with emotional, personalized attachment to man-hour calculation.

Moreover, teams assess story points collectively, whether that item is their domain (development, test, etc.) or not, discuss it briefly, and each member will mentally formulate an estimate. Next, every team member, at the same time, holds up a card revealing the number that reflects their estimate. If there are differences in the estimates, a discussion ensues until a consensus is reached.

This is called poker planning; it reinforces the objective method of estimation—not by an individual skill level.

3.4.7 Features and Epics

Tasks are the most granular building blocks of a user story. Three other types are important for requirements traceability, although they are not always used (**figure 3.22**).

- **Epics**—These can be thought of as the big ideas. Epics cannot be completed within a single iteration and are usually broken down to several features and user stories or incremental chunks of functionality.
- **Features**—A collection of user stories.
- **Tasks**—The to-do items that the owner(s) of the user story will perform, similar to a checklist.

Figure 3.22—Agile Requirements Model

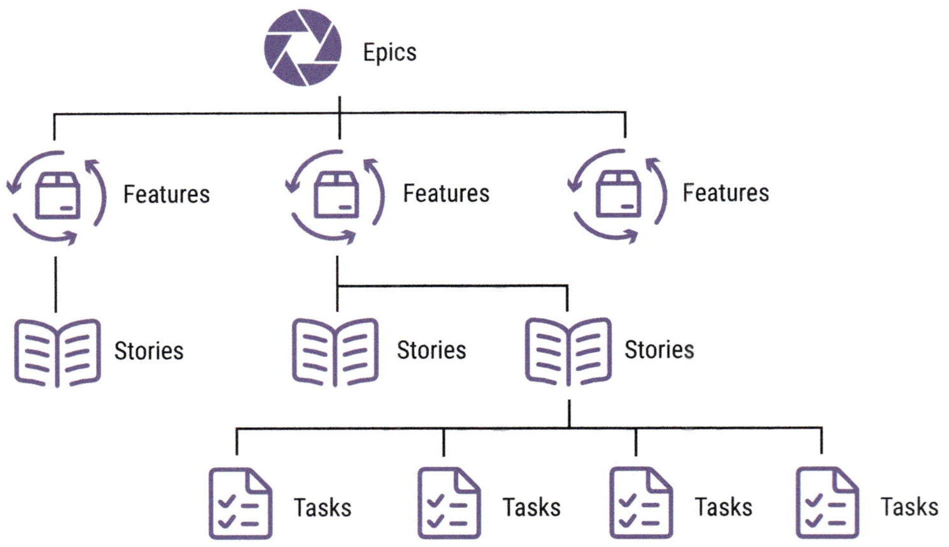

3.4.8 Planning and Schedules

- **Sprint**—The sprint is a time box schedule of one month or less during which the team produces a potentially shipable product increment. Typical characteristics of sprints are:
  - Sprint start date and end date are fixed
  - Development effort is consistent throughout sprint and is measured via a velocity chart. The scrum master provides velocity and burndown charts (see **figures 3.23** and **3.24**).
  - A new sprint immediately follows the conclusion of the previous sprint.
- **Sprint planning**—Sprint zero, also called sprint planning is the first iteration where the team discusses the backlog and decides which user stories to allocate for the first iteration. After the user stories are chosen, they are further refined by adding team-member tasks to the respective user stories. Each additional sprint includes a sprint planning meeting where the user stories are groomed for details and inclusion into that sprint.

  After the team and product owner establish the scope of the sprint, no more items can be added to the sprint backlog. This protects the team from scope creep changes within that sprint.
- **Daily standups**—The daily standup is a short 15-minute daily meeting (huddle) during which the team remains standing in a common area while each team member discusses progress on their tasks from the previous day and any blockages they may have encountered. The standup is not intended to be a status reporting meeting or a problem-solving discussion; it only raises a flag signaling a possible issue that may prevent the team from completing the sprint so it can be removed or mitigated quickly.
- **Sprint review**—At the end of the sprint, the entire team reviews the results of the sprint with stakeholders of the product. The purpose of this discussion is to discuss, demonstrate and potentially give the stakeholders a chance to use the completed features in that sprint, to get their feedback. The sprint review is not intended to provide a status report. Feedback from the sprint review gets placed into the product backlog for future consideration.

Agile allows the time that is usually spent on long project plans, documentation and meetings in the traditional project management lifecycle to be returned to the team.

3.4.9 Burndown and Velocity Charts

Velocity is defined with respect to units of value (user stories and their points) rather than with respect to units of effort (tasks and their corresponding hours). It is an indication of the *throughput of value* that is produced during a given sprint. The summation occurs at the end of each iteration, the team adds up effort estimates associated with user stories' points that were completed during that iteration. An example is depicted in **figure 3.23**. The lighter-shaded area indicates what is yet to be completed and the darker-shaded area indicates completed stories.

This allows the team to estimate or gauge how long it will take to complete future-sprint user stories by looking at averages of prior sprints. This velocity chart shows the team committed to more points than they have in the past (e.g., Sprint 9B was 195 points and Sprint 10A is 221.)

Burndown is the amount of work completed and what remains for a given sprint. A burndown chart is different than velocity in that it measures man hours, not user story points.

The burndown chart shaded area allows for a quick visual of any scope creep; if new user stories are added or removed within a sprint, it will show peaks and valleys flagging scope changes. The goal is for the shaded area and the slanted line to meet at the end of the chart to indicate that what was allocated was indeed completed on time.

Figure 3.23 depicts a sample velocity chart and **figure 3.24** is a sample burndown chart.

Figure 3.23—Sample Velocity Chart

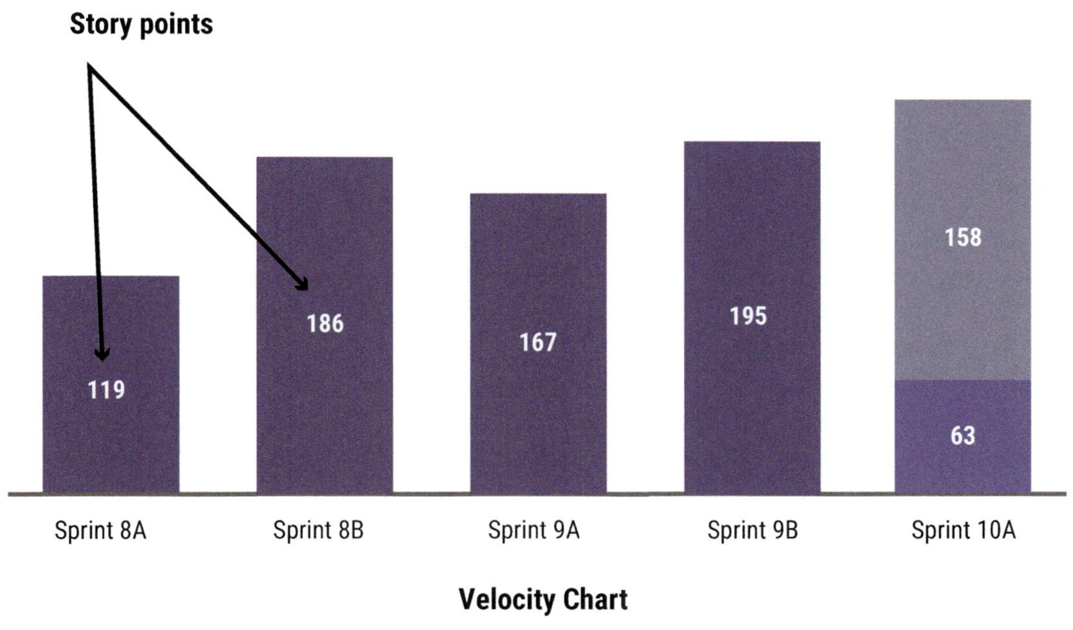

Figure 3.24—Sample Burndown Chart

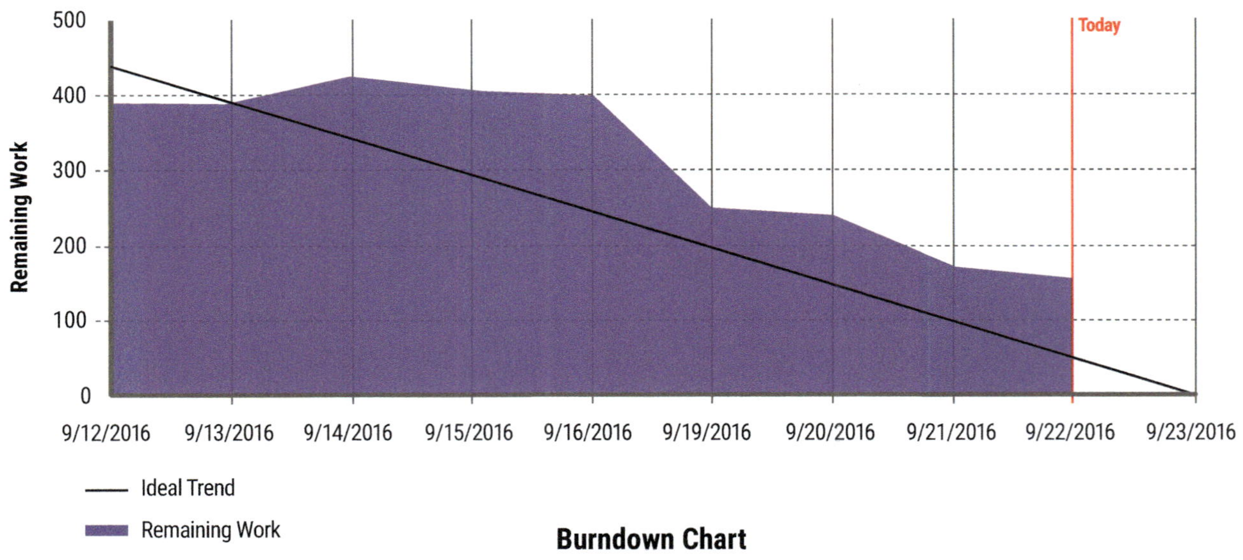

3.4.10 Scrum

Scrum is now the leading Agile development method used by Fortune 500 enterprises and others around the world.[54] The scrum values and principals on which the scrum framework is based are as follows:

Scrum Principals

- **Transparency**—Means presenting the facts as they are. All individuals on the team are involved in sharing information openly, regardless of hierarchical authority or roles. The CEO, customer and individual contributors are transparent in their day-to-day dealings with others. They share good and bad news alike with courage, striving for the common organizational objective.
- **Inspection**—This is not an inspection by an auditor, but rather self-checking and peer inspection by others on the scrum team. Inspections may include the product, processes, people and relationship aspects, best practices, and continuous improvements. For example, the team displays the product at the end of each sprint to the customer to gather valuable feedback. If recommendations or changes are requested during the sprint review, the team does not complain, but, rather, adapts by using this as an opportunity to collaborate, clarify requirements and test new hypotheses. The Agile approach is "expect the unexpected," where deviations are exploited as possible features and opportunities for innovation.
- **Adaptation**—The ability to adapt, or do things differently based on the results of the inspection. To continuously improve, everyone must ask regularly: Are we better off than yesterday?

Scrum Values

- **Commitment**—Scrum teams commit to achieving the sprint goal and to encourage each other during the sprint progression. If the team fails to meet the sprint goal, blame is not assigned to an individual—the scrum team is collectively accountable. Moreover, the team safeguards its commitment by flagging impediments (also known

[54] Scrum Alliance, "Why is it called Scrum?," 2014, www.scrumalliance.org/ScrumRedesignDEVSite/media/ScrumAllianceMedia/Public%20Relations_2/What-is-Scrum-Backgrounder-2014.pdf

as blocks) early, assisting each other with work, and ignoring day-to-day distractions. These practices produce dependability and further reinforce trust in each other.

- **Courage**—Courage is employed by the team each time the team asks clarification of ambiguity, chooses to ignore distractions, talks openly and respectfully about challenges, and holds self and each other accountable.
- **Focus**—Scrum teams focus on meaningful objectives that benefit the organization and serve the customer; they commit to the sprint goal and to delivering value in every sprint.
- **Openness**—Similar to transparency, open scrum teams use courage to bring problems to the table and speak about them respectfully. If things are unclear, open teams will be courageous in asking lots of questions, bring up difficult subjects and discuss challenges when needed.
- **Respect**—Respect for the individual and a belief that everyone is doing the best they can. Through respect, teams can be open with each other without fear of attack. They can ask for and give help. They can talk about challenges and challenge each other. They can be transparent, and they can depend on each other.

3.4.11 SCRUM Roles and Responsibilities

The following are SCRUM roles and corresponding responsibilities:

- **Product owner and stakeholders**—The product owner is the voice of the customer (i.e., the stakeholders). They are responsible for gathering requirements, creating user stories (i.e., requirements), managing, and prioritizing the product backlog (a repository where the user stories reside). Any prioritization or user story conflicting direction, or no direction at all with respect to what to build are presented to the product owner.
- **Scrum master**—The scrum master is similar to a project manager on traditional projects. In Agile, however, their responsibility is ensuring that the team lives Agile values and principles and follows the processes and practices that the team agreed they would use. Normally, the scrum master also facilitates all standups and meetings and provides statistics such as velocity and burndown charts.

  Moreover, given the rugby reference, the scrum master holds no actual authority and are considered coaches leading from a position of influence, often taking a servant-leadership stance.
- **Development team**—The development team consists of all of the people who deliver the product. That includes the developers, product owners, testers and scrum masters. In Agile, developers are equal to the rest of the team and each member is personally accountable to the success of the project.

  The main responsibility of the development team is to deliver the incremental value; how the work is divided is left up to the team to determine—it is a decentralized approach.

3.4.12 Extreme Programming (XP)

Extreme programming (XP) is one of three Agile software development frameworks that aims to produce higher quality software and higher quality of life for the development team. XP and SCRUM differ slightly in their application of the Agile value system. Five specific values are part of every XP team: communication, simplicity, feedback, courage and respect.

Within XP, several new techniques were created that can be leveraged, regardless of the Agile methodology selected. XP relies heavily on customers being available onsite, next to a small group of developers (2-12) for expedited access and feedback, thus making this approach more difficult to scale. These XP techniques are as follows:

- **Refactoring**—Agile is fast-moving development, so code writing and design may be hurried compared to waterfall. Refactoring allows a review of previously written, working code and improving its organization, design, comments, etc. Said differently, "Refactoring consists of improving the internal structure of an existing program's source code, while preserving its external behavior."[55] One pitfall about refactoring is that it can introduce new bugs as the code is rewritten or reorganized. Some argue that refactoring provides no customer facing value when compared to time spent on new features.

- **Pair programming**—Also known as pairing, involves two programmers working together on one piece of code. They sit next to each other and provide input and examination, resulting in built-in quality, information sharing and team building. The roles are sometimes referred to as driver and navigator, and the expectation is that the programmers swap roles often.

Figure 3.25 shows the short, constant feedback loops common in Agile and especially XP methodology.

Figure 3.25—Agile Constant Feedback Loop

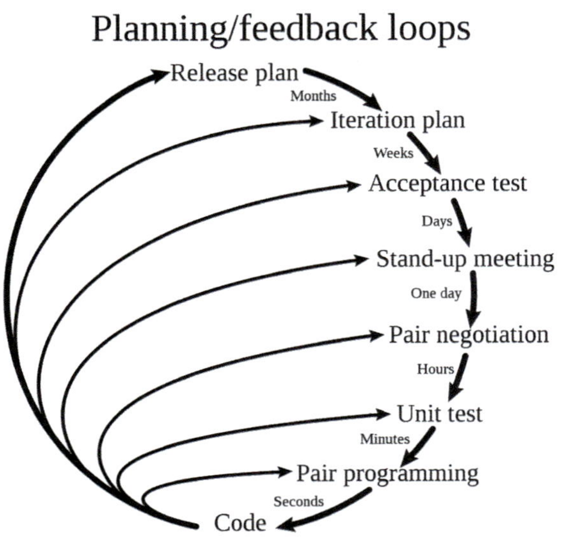

Source: Wells, J.D; "Extreme Programming (XP) requires planning and feedback at many levels and many frequencies," 26 July 2013, https://commons.wikimedia.org/w/index.php?curid=27448045. The figure is available under the Creative Commons Attribution 3.0 Unported license.

3.4.13 Feature-Driven Development (FDD)

Feature-driven development (FDD) was introduced in 1997 by Jeff De Luca when he was working on a software development project for a large Singapore bank. FDD was built around software engineering best practices, such as domain object modeling, developing by feature and code ownership. The blending of these practices that resulted in a cohesive whole is the best characteristic of FDD.[56]

FDD delivers just enough design initially (JEDI) functional software in a consistent timely manner, when contrasted with XP, which may be more unpredictable and volatile. FDD uses two-week sprints to achieve a set of working features that are granular enough to complete within that two-week period.

[55] Agile Alliance, "Refactoring," *Agile Glossary*, https://www.agilealliance.org/agile101/agile-glossary/#r
[56] Santos, J.; "XP, FDD, DSDM, and Crystal Methods of Agile Development," 2 October 2018, https://project-management.com/xp-fdd-dsdm-and-crystal-methods-of-agile-development/

Figure 3.26 shows the feature-driven development approach and the central idea of features to this methodology.

Figure 3.26—Feature Drive Development Approach

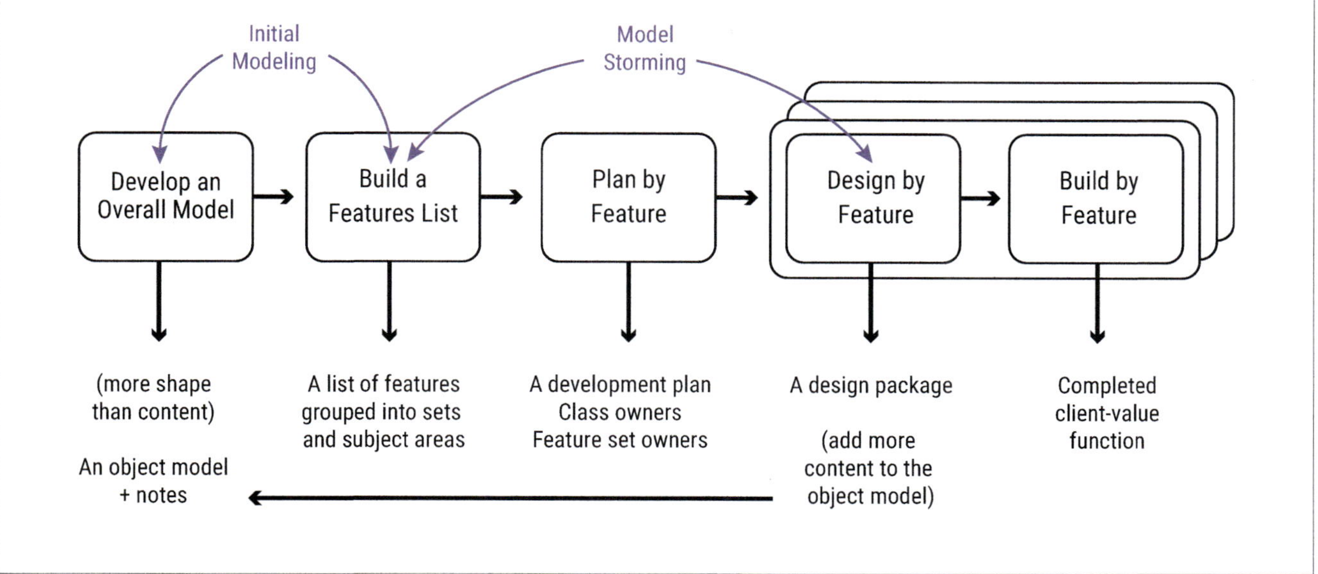

Source: Software Architectural Design in Agile Environments, 2018, https://www.researchgate.net/figure/Feature-driven-development-13_fig2_322192112. This figure is available under the Creative Commons Attribution 4.0 International license.

3.4.14 Choosing a Methodology

Selecting the right methodology for a development project is highly dependent on the project size and complexity, company industry and team culture. It is not uncommon for teams to leverage and piece-meal techniques from some methodologies, while choosing procedures and practices from another methodology, creating a hybrid approach. **Figure 3.27** outlines, at a high level, the major differences between the methodologies. It is worth noting that in some cases the terms Agile and Lean are interchangeable, and command responsiveness, adaptability, and continuous change and learning.

Figure 3.27—Software Development Methodologies Comparison

	Team	Time	Artifacts	Ceremonies	Role Titles	Testing
Waterfall	Customer offsite Managed team	Long cycles Can take years	• SOPs • Requirements documents • Traceability documents • Project charter sign off • User manuals • Risk register	• Kick-off meeting • Post mortem • Risk meetings	• Project manager • Program manager	• Manual • Regression testing • Test plans • Test cases
Scrum	Customer onsite Self-organizing team	30-day sprints range from 1-to-4 weeks	• Product backlog • User stories • Burndown charts • Velocity chart	• Iteration 0 • Daily scrum meetings • Sprint review • Sprint retrospective • Inspections • Refactoring • Code walkthroughs	• Scrum master • Scrum team • Product owner	• Manual and automated • Design inspection • Code review • Unit testing • Continuous integration
XP		Sprints range from 1 to 6 weeks				
FDD		Sprints range from 1 to 2 weeks	• Plan and design by feature	• Design inspection • Code review		

Note: In Agile methodologies, risk management does not have to involve the formal documentation and meetings of traditional waterfall methodology. Instead, risk management is built into scrum roles, sprints and ceremonies. Typically, as any project matures and progresses, the risk of that project generally declines regardless of methodology.

3.4.15 Versioning and Repositories

Software configuration management (SCM) is the practice of source code management with the goals of establishing best practices for build management, version control, and the foundation of baselines with an overarching process optimization.

Software versioning is a way to categorize the states of software as it is developed and released. The version identifier can be a word, a number, or both that will usually correlate to the progress of the application. In some cases, there is a correlation between versioning and software quality. For example, version 1.0 is commonly used to denote the initial release of a program, including noncomputer related releases.

A version control system (VCS) is used by software developers during a software development project with the purpose of maintaining versions and collaborative work. It is a part of configuration management.[57]

One such example is the well-known GitHub application. GitHub is an online, cloud-based VCS development platform used by over two-million top tier enterprises and 50-million developers.[58] The platform is subscription-

[57] Elsen, R.; I. Liem; S. Akbar; "Software versioning quality parameters: Automated assessment tools based on the parameters," 2016 International Conference on Data and Software Engineering (ICoDSE), Denpasar, 2016, pp. 1-6

[58] GitHub, "Where the world builds software, " https://github.com/

based (like most SaaS offerings) and offers different tiers (some open source) of memberships that cater to individual contributors, small and large businesses, and institutions. Other VCS examples are Microsoft Azure DevOps (also known as Team Foundation Server), BitBucket and SourceForge.

Software versioning is critical, considering how applications are used and the number of contributing developers working on the code. The version number is how users reference a unique release of an application that may have many variations in the market.

Software versioning is particularly important when considering:

- Software upgrade compatibility and precedence
- Reporting bugs and errors, particularly security flaws
- Collaborating with other developers

3.4.16 Software Versioning Schemes

A variety of version numbering schemes are available to keep track of different versions of a piece of software. Some are done with simple date and time to remain unique; other approaches use a sophisticated approach to categorize and reference. The omnipresence of computers has also led to these arrangements being used in contexts outside of computing.

SemVer Numerical Scheme

The semantic naming convention, better known as SemVer, is the most commonly used scheme. Each position of the version number indicates a type of release. Positions are separated with a decimal point. The first position indicates the major version number; the second position indicates the minor version number; and the third position indicates the patch number.

Example:

Major.Minor.Patch

1 .0 .0

1.0.0 notates a major release and is typically changed far less often for mature software.

1.1.0 notates a minor release update which may include some additional features.

1.1.1 notates bug fix(es) and may change often in early stages of software development.

Each place holder is also referred to as a revision number (i.e., revision control). The version numbers are usually assigned in increasing order and correspond to new changes in the software. Further detail reveals revision control is often used for recording and categorizing incrementally different versions of information, whether or not this information is computer software. SemVer is also used to denote documentation or regulation updates.

Alphabetical Naming Conventions

Naming conventions, such as Alpha, Beta, and RC, are some examples, although they can vary somewhat depending on enterprise nomenclature.

- Alpha version refers to the first release of a program.

- Beta version refers to the second release of a program.
- RC is short for release candidate version of a program.

Alpha and Beta may be released internally only or to a limited set of customers that volunteer to test their vendor's updates or requested features. A release candidate will be released with the semantic versioning discussed above.

Precedence refers to how versions are compared to each other when placed in numerical order. Precedence *must* be calculated by separating the version into major, minor, patch and pre-release identifiers, in that order.[59]

Precedence is determined by the first difference when comparing each of these identifiers from left to right as follows: Major, minor, and patch versions are always compared numerically.[60]

Example:

1.0.0 < 2.0.0 < 2.1.0 < 2.1.1.

Numeric identifiers always have lower precedence than nonnumeric identifiers. A larger set of pre-release fields has a higher precedence than a smaller set, if all of the preceding identifiers are equal.[61] In some cases, a tag may also be appended to the version number; a tag is a textual label further identifying that version. For example:

1.0.0-alpha < 1.0.0-alpha.1 < 1.0.0-alpha.beta < 1.0.0-beta < 1.0.0-beta.2 < 1.0.0-beta.11 < 1.0.0-rc.1 < 1.0.0.

Software projects are created in a code repository, also known as code repo. A repository is a workspace used by all developers to organize, collaborate, control versioning, submit and check out files, test, back up, and archive source code (**figure 3.28**).

[59] "Semantic versioning 2.0.0," https://semver.org/

[60] *Ibid.*

[61] *Ibid.*

Figure 3.28—Repository Architecture

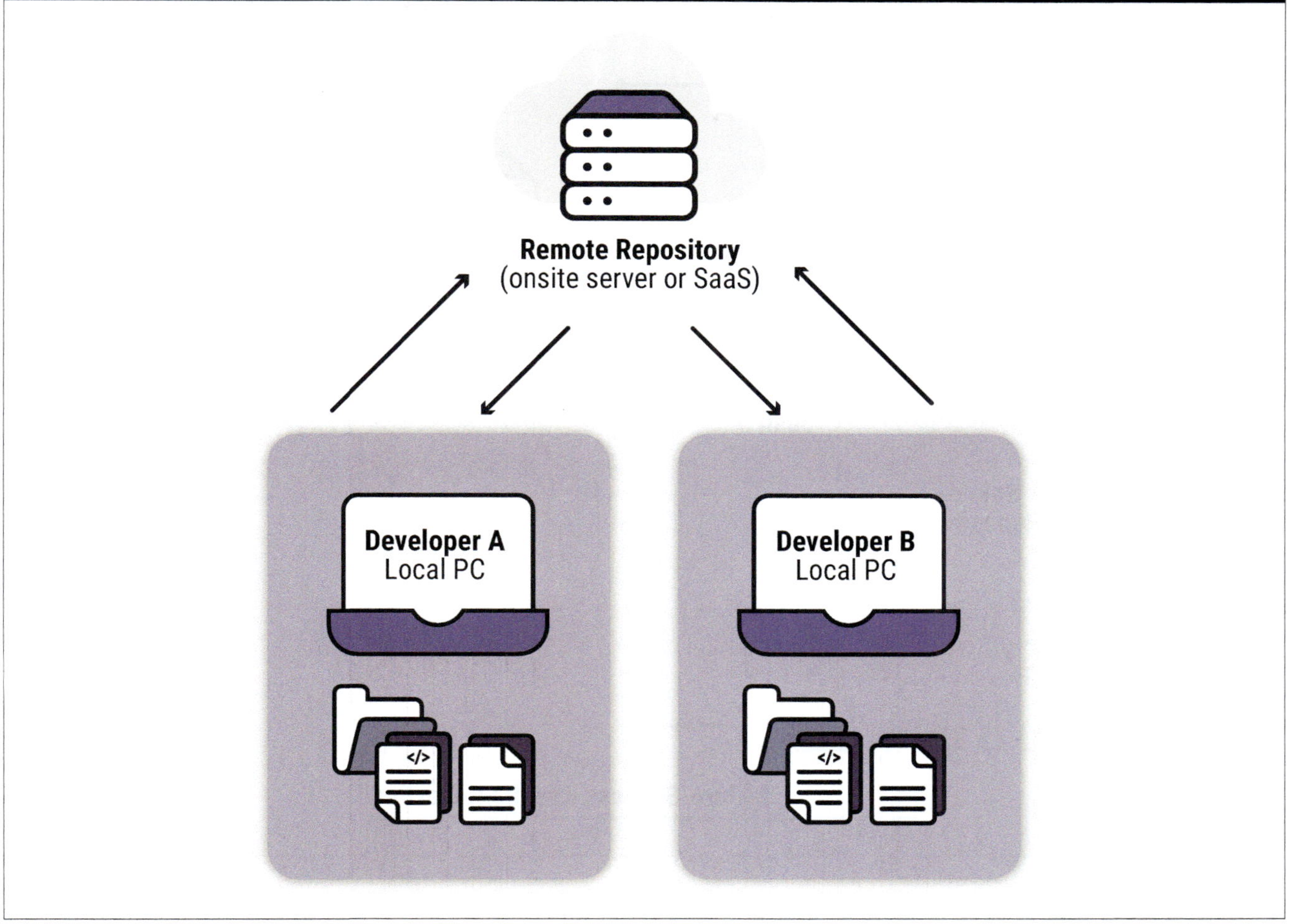

Individual developers can work on separate files and check them in when complete. This shared repository prevents redundancy and manages overhead that would otherwise be required by each developer locally. Overhead tasks, such as back-up, reporting, version control, logging, management, testing and organization, are now mitigated and automated.

3.4.17 Repository Mechanics

The repository is a collaborative environment with continuous built-in integration, as seen in **figure 3.29**. It enables programmers to perform the following critical tasks (some are done by the repository automatically):

- Check-out code: also known as fetch
- Check-in code: also known as push
- Trigger pull requests: the code that is checked in gets peer reviewed prior to commitment
- Commit code: senior or lead developer commits code post peer review and discussion
- Trigger a build
- View changelogs
- Rollback changes

Figure 3.29—Repository Architecture and Functionality

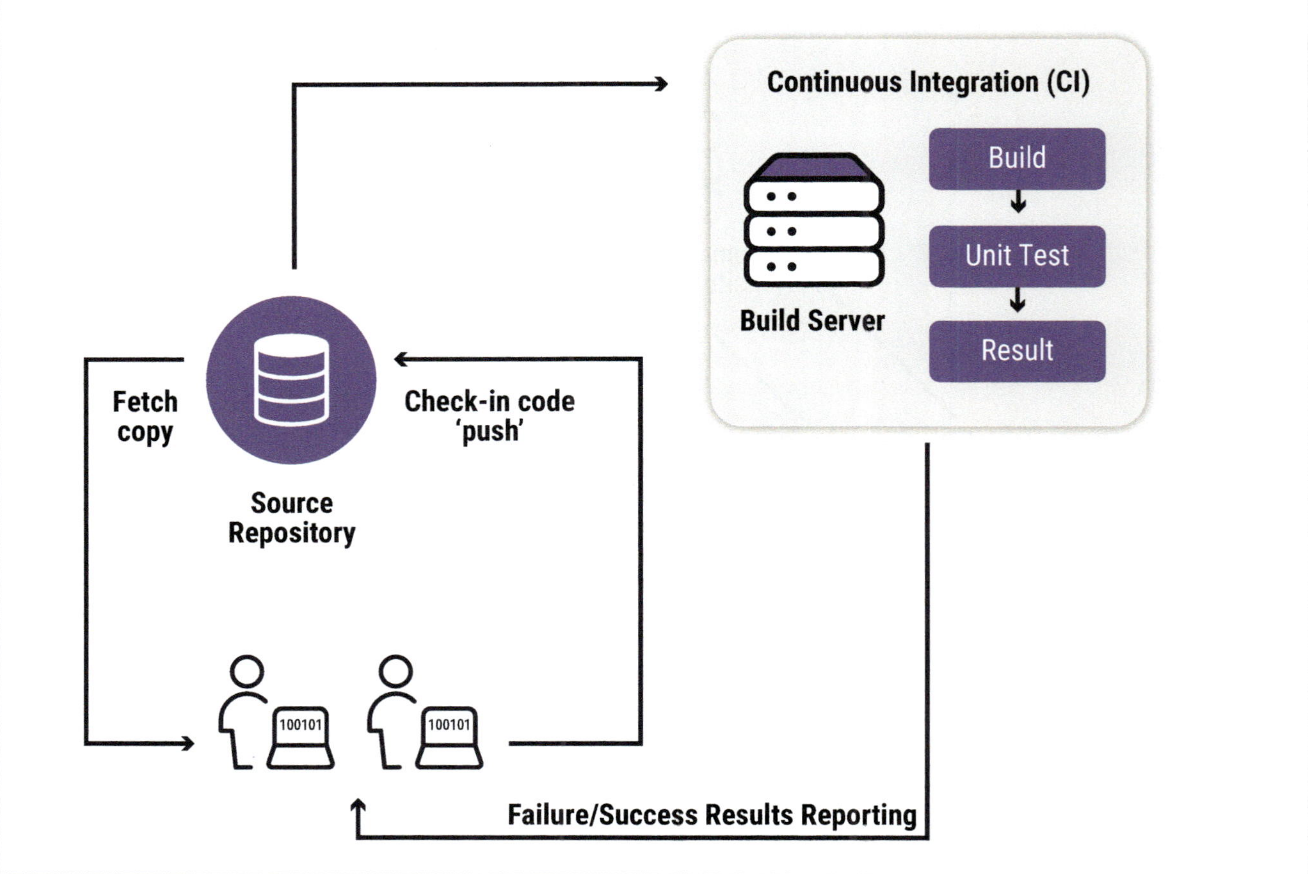

3.4.18 Repository Code Branches Structure

Branching, in version control and software configuration management, is the duplication of an object under version control (such as a source code file or a directory tree) so that modifications can occur in parallel, along multiple branches. Branches are also known as trees, streams or codelines.[62]

The source code repository is comprised of several branches that dictate the pathways of leading away from, and towards, the mainline. There are several patterns that can allow teams to use branching effectively in an organized manner to the production releases.

The originating branch is sometimes called the master branch, the upstream branch (or simply upstream, especially if the branches are maintained by different organizations or individuals), or the backing or parent stream. Child branches are branches that have a parent; a branch without a parent is referred to as the trunk or the mainline.[63]

An overview of a common practice scheme of a software configuration repository is shown in **figure 3.30**.

- **Master**—As production branch
- **Develop**—As a development branch
- **Feature**—As a prefix for all feature branches

[62] Wikipedia, "Branching (version control)," https://en.wikipedia.org/wiki/Branching_(version_control)
[63] Berczuk, S.; B. Appleton; *Software Configuration Management Patterns: Effective Teamwork, Practical Integration*, Addison-Wesley, 2003

- **Release**—As a prefix for all release branches
- **Hotfix**—As a prefix for all quick release of bug fix branches

Figure 3.30—Sample Scheme for Software Configuration Repository

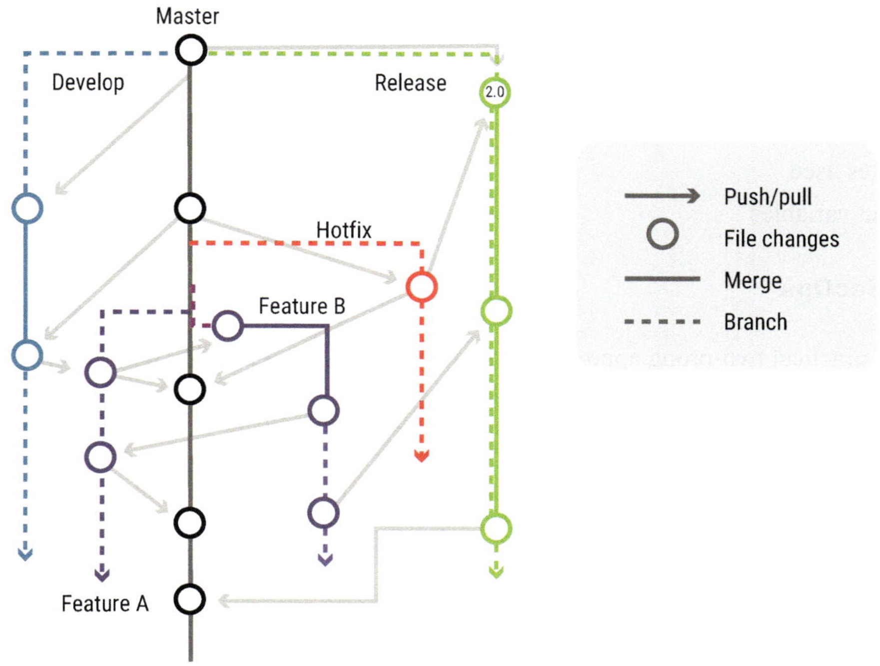

Source: Adapted from Dagher, F.; "iPaaS 12 – Branching Strategy," https://fadidagher.net/2018/02/26/ipaas-12-branching-strategy/

- **Master**—This is the baseline of the code without direct access for the developers. The purpose of the Master single branch is to keep the most stable deployable version of the code. No one develops directly on this branch.

  Management must ensure that development staff members cannot modify production programs or applications. The software engineers, programmers, developers or coders should work in a test-only environment and follow check-in procedures or turn their work over to another group to move programs and application changes into the production environment in an organized, traceable manner to reduce risk.[64]

- **Develop**—This is the main focus point for the developers where all active development takes place, Feature branches are created from and later merged to this branch.
- **Hotfix**—These are categorized as bug-fixing branches. When a new bug is discovered on a released version, a new hotfix branch is created from the Master. After fixing it, it is merged back to the Master and to Develop. This is to ensure that the released bug fix carries forward to any future development efforts that occur in the Develop branch.
- **Feature**—A set of branches categorized as Feature branches. Here, new functionalities and features are created. Feature branches are always derived from a Develop branch and are always merged back with it.
- **Release**—When the time comes to release a new version of the app, this is the place from which a build is triggered and includes all previously merged code into this branch.

The immense benefit of having a centralized repository with this scheme is that each time a push and then a merge occurs from any input branch back to its parent branch, the source code is compiled as a whole, and tested via automated tests called unit tests.

[64] *Op cit* ISACA, *CISA Review Manual, 27th Edition*

This process is commonly known as continuous integration and is started when a build request is triggered from the release branch.

Continuous integration (CI) means that source code is continuously being built into binary artifacts that reside on a build server as shown in **figure 3.30**. Proper integration with continuous integration servers provides several metrics that assist should a build request fail, such as:

- Which user triggered the build
- Which modules were built
- Which sources were used (commit id, revision, branch)
- Dependencies used
- Environment variables

3.4.19 DevSecOps

DevSecOps is a practical two-prong approach to integrating information technology security into the applications created during the software development lifecycle. It is concerned both with the code algorithms and how they are written, ensuring no backdoor malicious capabilities. Backdoor is a means of regaining access to a compromised system by installing software or configuring existing software to enable remote access under attacker-defined conditions. The second consideration is it incorporates the business security requirements of using the application by the intended end users.

DevSecOps requires a collaborative working relationship between the SDLC team and the information technology security team of the business, to incorporate best practices, new requirements and test case validation as they relate to security only and not the feature-based business requirements of the application .

Several controls can be utilized throughout the SDLC, at different stages, to ensure that security is interwoven during the development process:

- **User management and application access control**—This is the first line of defense to prohibiting unauthorized access onto the application. User provisioning can be done at the application level only or integrated into an organizational-wide, federated single sign-on (SSO) that is initiated as part of the hiring process. Furthermore, it ensures that once the employment is terminated, the application access is also terminated.
- **Authorization levels**—Applications provide different feature sets and capabilities; therefore, a user's level of access to an application must match their job role. It is important to ensure that administrative or root level access (i.e., privileged account) is given to as few users as possible, if any, because that introduces a security risk. Privileged accounts should be monitored closely and reviewed regularly. API and automated-script access should also be thought of as users that can be used to cause havoc, if taken over by a backdoor attack, and should never have root level access.
- **Patch/hot fix management**—After software is released to the market, it is imperative for software to be updated regularly, whether for bug fixes or more so for security updates, because new threats are constantly surfacing. This is especially important for specific industries, such as banking and government agencies.
- **Configuration and change management**—These are responsible for ensuring proper release schedules, version control organization and that all change requests follow a defined, repeatable and documented process to ensure that all requested changes are fully assessed for any impact they may invite to either the software functionality or security of the application at all levels: information loss, access control changes, security breaches and the like.
- **Logging and auditing**—The business information technology security team, along with the application development team, should implement logging of all activities on the network in addition to application-specific logging and audit trail. This logging ensures both a proactive approach by using the controls and a reactive approach by reviewing logs for any suspicious activities.

3.4.20 Difference Between DevOps and DevSecOps

DevSecOps is different in its focus from DevOps in that DevSecOps is concerned with security only, and DevOps is concerned with application development speed and the quality of the features for the operational, functional side of the program. These competing priorities and requirements need to be considered, bearing in mind schedules and time to market, while weighing the risk of each decision.

3.4.21 Secure SDLC

The focus of secure SDLC is integrating cyber, application and business security into the SDLC in a more refined, phase-dependent, requirements-based approach. The secure SDLC model follows the standard model with a few additions and variations that are focused on continuous security integration and attention. Two distinct proposals are available today:

- **NIST SP 800-160 V1**—Published and observed by the US federal agencies for application development. The National Institute of Standards and Technology (NIST) developed this volume as a supplement to the existing, well established SDLC process approved by ISO and IEEE. The NIST standard takes into account a holistic organizational approach to the security of an individual application as it considers all input that could be a security threat from all layers of an organization.

Figure 3.31 illustrates the NIST security requirements analysis, showing the multitude of factors that should be considered.

Figure 3.31—NIST Security Requirements Analysis

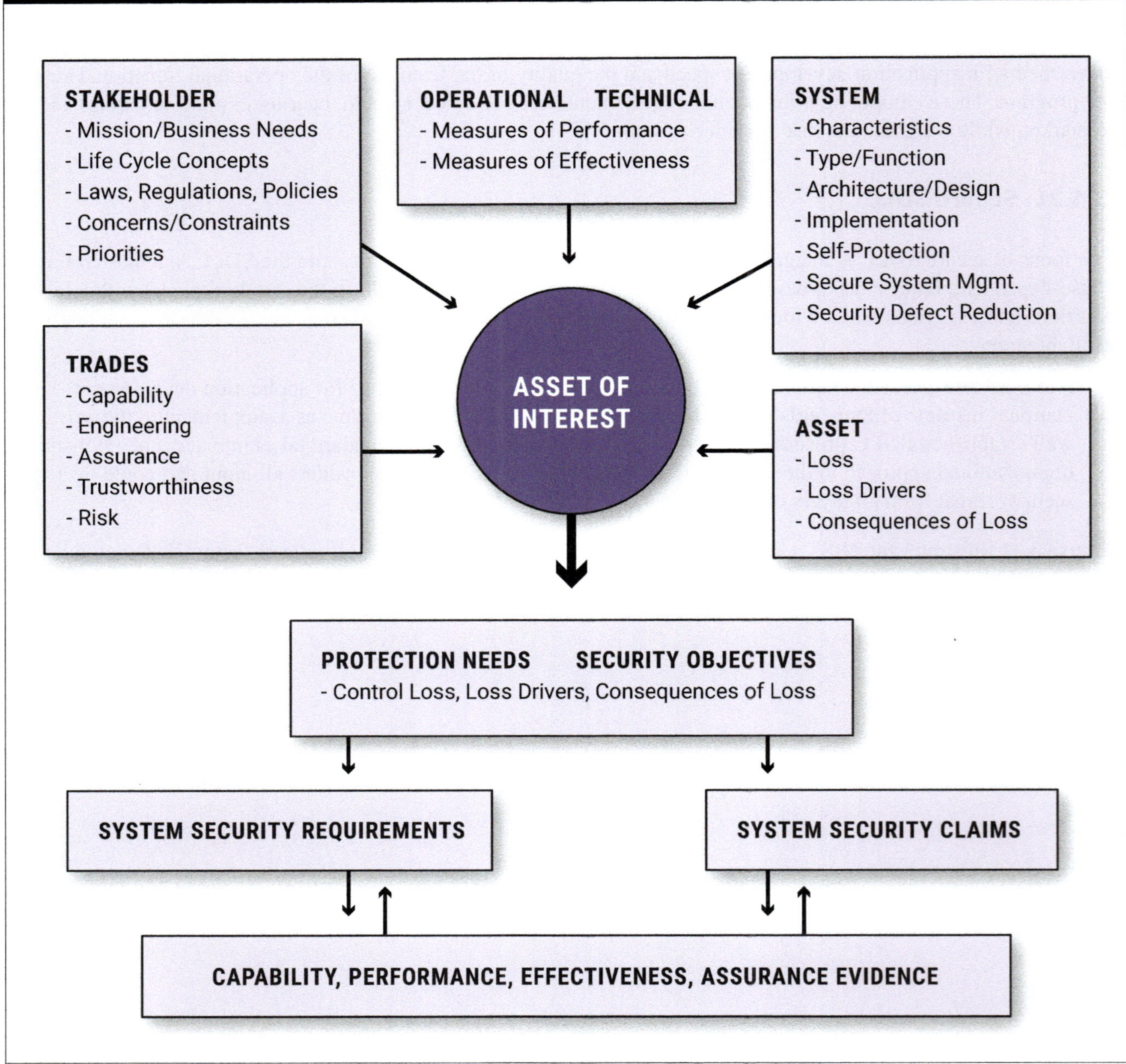

Source: Ross, Ron; Michael ; Janet Carrier Oren; "Systems Security Engineering: Considerations for a Multidisciplinary Approach in the Engineering of Trustworthy Secure Systems," NIST SP 800-160, Vol. 1, November 2016, fig. G-4, National Institute of Standards and Technology (NIST), https://nvlpubs.nist.gov/nistpubs/SpecialPublications/NIST.SP.800-160v1.pdf

- **Microsoft Security Development Lifecycle (MS SDL)**—Also known as Secure DevOps, MS SDL is a hybrid of the traditional Waterfall methodology with the iterative Agile model to form the spiral model of SDLC. The spiral model takes advantage of the systematic, controlled aspects of the Waterfall model with the iterative, quick to respond, short feedback model with very high emphasis on risk analysis. Microsoft SDL emphasizes specific practices to incorporate into each of the SDLC phases that are security focused. **Figure 3.32** illustrates those practices.[65]

[65] Microsoft, "What are the Microsoft SDL practices?," https://www.microsoft.com/en-us/securityengineering/sdl/practices

Figure 3.32—Microsoft SDL Practices for Software Development

Training

- Core security training for all stakeholders involved in the SDLC process

Requirements

- Define security requirements
- Define quality gates
- Define bug bars
- Define metrics and compliance reporting
- Assess security and privacy risk

Design

- Establish design requirements
- Analyze attack surface
- Perform threat modeling

Implementation

- Use approved tools
- Deprecate unsafe function
- Define and use cryptography standards

Verification

- Perform dynamic analysis
- Perform fuzz testing
- Perform attack surface review
- Perform static analysis security testing (SAST)
- Perform dynamic analysis security testing (DAST)
- Perform penetration testing

Release

- Plan and document incident response plan
- Perform final security review

Response

- Execute incident response plan

There are several testing techniques and development tools predominantly used for security in the SDLC in general:

- Penetration testing focuses on white-hat hacking to simulate attacks on web applications by identifying vulnerabilities in the network's firewall to access the application or via source code exploitation via full source code scanning.
- Code security plugins provided by Microsoft can be installed in Visual Studio IDE and alert the developer to any security flaws while the code is written; Roslyn analyzers (.NET Compiler Platform code analyzers that check for style, quality, maintainability, etc.) can be used during the build phase.
- DevOps specific tools assist with automated scripts and extensions to look for known security weakness proactively.
- Fuzz testing is an automated test that sends unexpected, invalid inputs to the application to uncover unhandled exceptions, crashes, failed code and memory leaks.

3.4.22 Project Management Tools and Techniques

The project management profession covers the breadth and depth of many domains. As project management continues to advance and improve, it offers many tools and techniques, and more are created every day. Following is a compilation of techniques that are standard protocol and are extremely useful, regardless of the industry, methodology or application developed. Note that the following tools and techniques do not represent the complete scope of project management.

Tools

Gantt Charts are one way that can be used to communicate project schedules. Gantt charts are a visual representation of time-bound activities as they relate to a project. They can be created in a simple spreadsheet program or in something as complex as Microsoft Project. These charts allow for organizing activities by phase or resources, compare planned versus actual completion dates, link work tasks as dependents and move schedules around as a whole, simply by shifting the start date of the project or task.

Most tools offer email notifications of upcoming due dates and the ability to share the schedule status via email in an automated cadence.

Figure 3.33 shows an example of a Gantt chart detailing the common project symbology.

Figure 3.33—Sample Gantt Chart

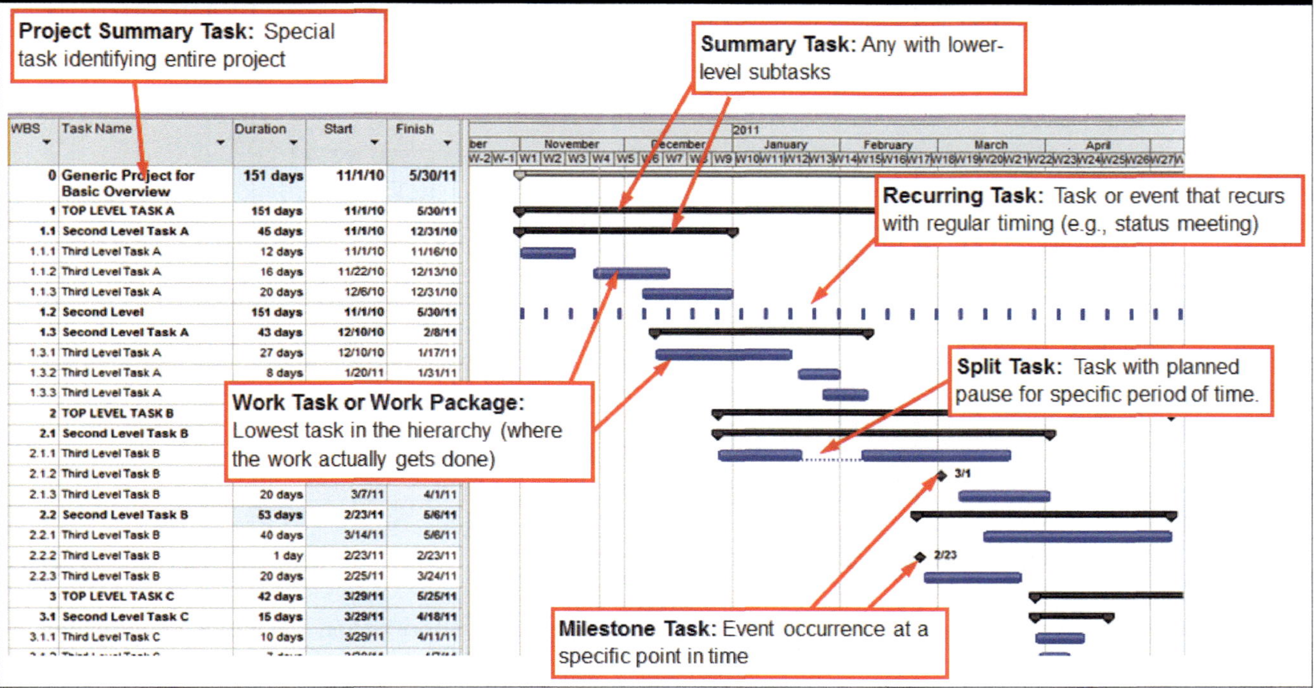

Source: Penn State College of Earth and Mineral Sciences, Department of Geography, https://www.e-education.psu.edu/geog871/l5_p5.html. This figure is available under Creative Commons Attribution NonCommercial-ShareAlike 4.0 International license.

Kanban is a tool *and* a technique recently embraced by traditional project managers when working on agile projects where time is of the essence, while detailed complex time schedules are reserved for waterfall-type projects (**figure 3.34**).

Kanban was created in the early 1940s by a Toyota car manufacturer Taiichi Ohno, to manage the planning and inventory status on the production line. Similarly, project managers may use a Kanban board as simple as sticky notes on a white board or electronically via a software application to visually display the current sprint status. The electronic version is especially useful for physically distributed teams.

There are variations to this Kanban board, although ultimately Kanban displays the following:

- To do
- Work in progress (WIP)
- Done

Testing and development groups may add *pass, fail and backlog* columns, and in some cases a *blocked* column is also added to draw attention to a blocked task. A Kanban board reinforces the transparency and communication values of Agile and is often used in the area where the weekly standups take place.

Figure 3.34—Sample Kanban Board

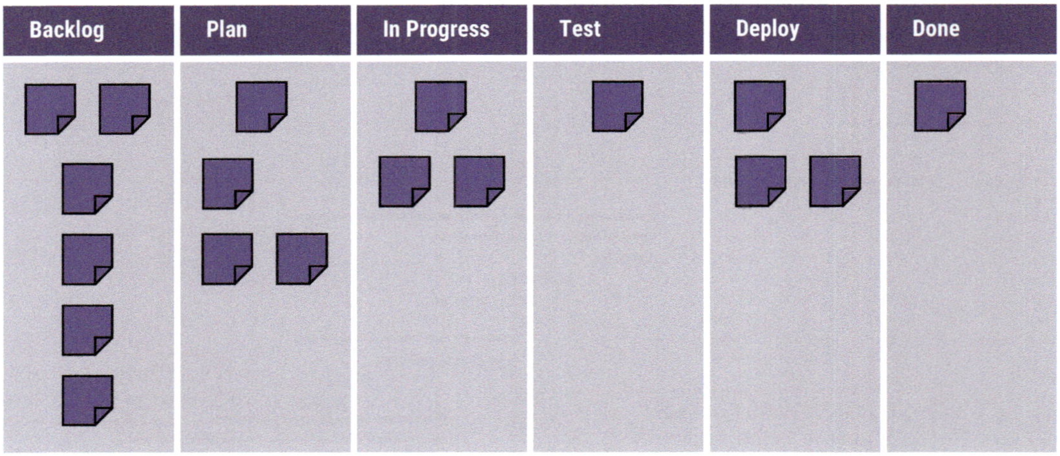

3.4.23 Time Management and Scheduling Techniques

One of project management's most coveted skills is the ability to enhance prioritization and provide information to the stakeholders that enables better decision making; WBS, PERT, and CPM are three such techniques.

Work breakdown structure (WBS) is a technique used to break down a very large project into manageable work tasks, or units, that are easier to track and manage and can help predict project schedules more accurately. WBS can be used for more than just actual project tasks. It is used for managing project budgets and finance, and resources, when working in a highly matrixed organization where resources have to be shared across different projects.

Work items can be identified as follows:

- **Float**—Work item has slack; opportunity to be pulled in or pushed out.
- **Free float**—Work item can be *delayed* without affecting the sequential task or interrupting a schedule.
- **Total float**—The amount of time that a scheduled activity can be *delayed* or *extended* without affecting the sequential task or interrupting a schedule.

WBS can be likened to the concept of epics in Agile, when it is broken down to its smallest parts called user stories. In Agile, user story points must be able to be completed within a sprint. Hence, the worst-case scenario is the sprint time block (usually one to two weeks). If the number of story points cannot be completed within two weeks, the story is still too large and must be divided into multiple user stories.

WBS is often used with program evaluation and review technique (PERT) for time estimation and prediction. PERT is a statistical approach to predicting project estimated completion time (known as E) while identifying a three-tier estimate: optimal delivery date (O), median (i.e., typical, most likely, M), and finally, worst-case (i.e., pessimistic, P) delivery date. The following is the PERT formula:

$E = (O + 4M + P) / 6$

Example: An enterprise is developing an application. The WBS is detailed, and every task is roughly estimated. The optimistic time estimate was 180 days (O), the most likely time estimate was 365 days (M) and the pessimistic time estimate was 600 days (P).

The equation using PERT:

$E = (180 + 4x\ 365 + 600) / 6$

Completed, this equation gives the project a PERT estimate of 373.33 days to finish all the work for this project.

Finally, the critical path method (CPM) is used to identify the length of time that each work path requires. The WBS that was created earlier is used as an input to the CPM calculation.

The time for each work path can be calculated using the longest path from the planned start date to the task completion date, including the earliest and latest time these tasks can start and finish without impacting the project schedule.

After the dates are collected, the tasks that are critical to the project will stand out as the critical path, either because those tasks have many dependencies or have no float available due to relying on other schedule and project constraints (e.g., the UI designer will not be hired until a certain date.)

Therefore, as a project management technique, CPM is a sequence of tasks that add up to the longest overall duration, whether there is float (whether free float, which is a dependent task, or total float, which is the overall project) or not. With this information, the shortest time needed to complete the project can be determined.

3.4.24 Computer Software Development Ethics

Given that software is used in most aspects of a person's life and interwoven in enterprise products and services, it is important to address the responsibility of the developer when writing programs and code that have real, tangible personal and social consequences.

There are several aspects of developer responsibility that should be considered, as outlined by the IEEE Software Engineering Code of Ethics and Professional Practice along with the Association for Computing Machinery (ACM) Code of Ethics and Professional Conduct.[66] Some of these are highlighted in the following paragraphs.

- **Give proper credit for intellectual property**—With the sharing and opportunities available on platforms, such as GitHub and other forums, ensure that intellectual property is attributed to the original author of the source code via comments in the code and/or help files.
- **Respect the privacy of others**—Cyberattacks are all too common and exploit systems to take advantage of a vulnerability; an example is a malicious PDF, crafted by the attacker and delivered by email. An adversarial threat event is made by a human threat agent (or adversary), while a nonadversarial threat event is usually the result of an error, malfunction or mishap of some sort[67] that could result from poorly written code.

  Only the minimum amount of personal information necessary should be collected in a system. The retention and disposal periods for that information should be clearly defined, enforced and communicated to data subjects. Retention and disposal periods should adhere to applicable regulations, including country laws and enterprise policies. Personal information gathered for a specific purpose should not be used for other purposes without the person's consent.

[66] CORPORATE IEEE-CS-ACM Joint Task Force On Software Engineering Ethics, "Software Engineering Code of Ethics and Professional Practice: Version 4," June 1998, https://dl.acm.org/doi/10.1145/276755.276772

[67] ISACA, *CSX Cybersecurity Fundamentals Study Guide, 3rd Edition*, USA, 2021

- **Identify, and address ethical, economic, cultural, legal and environmental issues related to work projects**—If a programmer notices and identifies that working on a project will lead to any kind of problems, then the programmer should report it before continuing. It is the developer's individual responsibility to report when programmers are exposed to be participants in illegal activities (either knowingly or unknowingly) to get money, fabricating or falsifying data, or censor information.

  People may get involved due to threats, economic issues or simply because they want to obtain monetary gain by taking advantage of their knowledge about how computer systems work. This guideline prohibits programmer involvement in such unlawful actions.

3.5 Chapter 3 Knowledge Check

REVIEW QUESTIONS

1. Entity relationship diagrams (ERDs) are created during which software development life cycle (SDLC) phase?
   A. Implementation
   B. Planning
   C. Design
   D. Testing

2. Choose the correct answer. In a sequential model, the customer can preview a subset of features prior to final release of the product.
   A. True
   B. False

3. Choose all that apply. Quality control (QC) is ________ oriented:
   A. Reactive
   B. Process
   C. Proactive
   D. Product

4. Choose the **BEST** option. The most cost-effective design option is:
   A. Paper prototyping
   B. Prototype
   C. Skeleton program
   D. Source code

5. Fill in the blank by entering the letter that corresponds to requirements characteristics.

Clear (concise, simple, defined) and understandable		A. To understand the requirement, there should not be a need to know any other requirement.
Unambiguous		B. A requirement is not necessary if it is not needed by any stakeholders, or if removing the requirement will not cause loss of functionality.
Testable (verifiable)		C. If a requirement contains facts, these facts should be true all the time; otherwise, each should be a separate requirement.
Correct		D. Requirements should not contain unnecessary verbiage or information.
Necessary and nonredundant		E. The test should either pass or fail.
Feasible (realistic, possible)		F. There should be only one way to interpret the requirement. All acronyms should be spelled out.
Independent and complete		G. The requirement should be doable within existing constraints, such as budget, schedule, and available resources.
Atomic		H. When in doubt break down the requirement into multiple one feature element.

Answers on page 144

Chapter 3 ANSWER KEY

Review Questions

1. A. Implementation
   B. Planning. Refer to page 96.
   C. Design
   D. Testing

2. A. True
   B. False. Refer to page 107.

3. **A. Reactive. Refer to page 109.**
   B. Process
   C. Proactive
   D. Product. Refer to page 101.

4. **A. Paper prototyping. Refer to page 105.**
   B. Prototype
   C. Skeleton program
   D. Source code

5. Fill in the blank by entering the letter that corresponds to requirements characteristics. **Refer to pages 98-99.**

Clear (concise, simple, defined) and understandable	**D**	A. To understand the requirement, there should not be a need to know any other requirement.
Unambiguous	**F**	B. A requirement is not necessary if it is not needed by any stakeholders, or if removing the requirement will not cause loss of functionality.
Testable (verifiable)	**E**	C. If a requirement contains facts, these facts should be true all the time; otherwise, each should be a separate requirement.
Correct	**C**	D. Requirements should not contain unnecessary verbiage or information.
Necessary and nonredundant	**B**	E. The test should either pass or fail.
Feasible (realistic, possible)	**G**	F. There should be only one way to interpret the requirement. All acronyms should be spelled out.
Independent and complete	**A**	G. The requirement should be doable within existing constraints, such as budget, schedule, and available resources.
Atomic	**H**	H. When in doubt break down the requirement into multiple one feature element.

Glossary

A

Address—1. A number, character or group of characters that identifies a given device or a storage location, which may contain data or a program step.

2. To refer to a device or storage location by an identifying number, character or group of characters.

Addressing exception—An exception that occurs when a program calculates an address that is outside the bounds of the storage that is available to the program

See Unhandled exception.

Algorithm—A finite set of well-defined, unambiguous rules for the solution of a problem in a finite number of steps, it is a sequence of operational actions that lead to a desired goal and is the basic building block of a program

Algorithm analysis—A software verification and validation (V&V) task to ensure that the algorithms selected are correct, appropriate and stable, and meet all accuracy, timing and sizing requirements

American National Standards Institute (ANSI)—The organization that coordinates the development of US voluntary national standards for nearly all industries. It is the US member body to the International Organization for Standardization (ISO) and the International Electrotechnical Commission (IEC). Information-technology industry standards pertain to programming languages, electronic data interchange, telecommunications and physical properties of diskettes, cartridges and magnetic tapes.

Analysis—1. To separate into elemental parts or basic principles to determine the nature of the whole 2. A course of reasoning showing that a certain result is a consequence of assumed premises 3. The methodical investigation of a problem and the separation of the problem into smaller related units for further detailed study (Source: ANSI)

Application programming interface (API)—A set of routines, protocols and tools referred to as building blocks used in business application software development

Application software—Software designed to fill specific needs of a user; for example, software for navigation, payroll or process control. Contrasts with support software and system software.

Architectural design—1. The process of defining a collection of hardware and software components and their interfaces to establish the framework for the development of a computer system. See Functional design.

2. The result of the process in definition 1 See Software engineering.

Archive—A lasting collection of computer system data or other records that are in long term storage

Arithmetic logic unit (ALU)—The area of the central processing unit that performs mathematical and analytical operations

Array—An n-dimensional ordered set of data items identified by a single name and one or more indices, so that each element of the set is individually addressable, e.g., a matrix, table or vector

ASCII—The American Standard Code for Information Interchange (ASCII). Uses 7 or 8 bits to represent an alphanumeric symbol or special character.

Assembler—A computer program that translates programs (source-code files) that are written in assembly language into their machine-language equivalents (object-code files). Contrasts with compiler and interpreter.

See Cross-assembler, Cross-compiler.

Assembly Language—A low-level programming language that corresponds closely to the instruction set of a computer, allows symbolic naming of operations and addresses, and usually results in a one-to-one translation of program instructions (mnemonics) into machine instructions.

Asynchronous transmission—Character-at-a-time transmission.

Audit trail—Data in the form of a logical path linking a sequence of events, used to trace the transactions that have affected the contents of a record

Source : ISO

Auxiliary storage—Storage device other than main memory (RAM), e.g., disks and tapes

B

Baseline—A specification or product that has been formally reviewed and agreed on, serves as the basis for further development, and can be changed only through formal change control procedures

BASIC—Beginners All-purpose Symbolic Instruction Code (BASIC) is a high-level programming language intended to facilitate learning to program in an interactive environment.

Benchmark—A standard against which measurements or comparisons can be made

Binary—The base 2 number system (2^n). Permissible digits are 0 and 1.

BIOS—Basic input/output system

Bit—A contraction of the term binary digit, and the most basic and smallest unit of computing information. A bit may be in one of two states, logic 1 or logic 0. It can be thought of as a switch that is either on or off. Bits are usually combined into computer words of various sizes, named bytes.

Blueprint—An exact or detailed plan or outline

Bomb—A Trojan horse that attacks a computer system when a specific logical event occurs (logic bomb) or when a specific time-related logical event occurs (time bomb), or is hidden in electronic mail or data and triggers a computer system attack when read in a certain way (letter bomb)

Similar to: Trojan horse, virus and worm

Boolean—Pertaining to the principles of mathematical logic developed by George Boole, a nineteenth century mathematician. Boolean algebra is the study of operations carried out on variables that can have only one of two possible values, i.e., 1 (true) and 0 (false). Like add, subtract, multiply and divide are the primary operations of arithmetic, and, or and not are the primary operations of Boolean Logic. In Pascal a Boolean variable is a variable that can have one of two possible values, true or false.

Boot—1. To initialize a computer system by clearing memory and reloading the operating system

2. To cause a computer system to reach a known beginning state. A boot program, in firmware, typically performs the boot function, which includes loading basic instructions that tell the computer how to load programs into memory and how to begin executing those programs. A distinction can be made between a warm boot and a cold boot. A cold boot starts the system from a powered-down state. A warm boot restarts the computer while it is powered up. Important differences between the two procedures are:

- A power-up self-test, in which various portions of the hardware, e.g., memory, are tested for proper operation, is performed during a cold boot, while a warm boot does not normally perform such self-tests.
- A warm boot does not clear all memory.

Bootstrap—A short computer program that is permanently resident or easily loaded into a computer, and whose execution brings a larger program, such as an operating system or its loader, into memory

Boundary value—1. A data value that corresponds to a minimum or maximum input, internal or output value specified for a system or component

2. A value that lies at, just inside or just outside a specified range of valid input and output values

Boundary value analysis—A selection technique in which test data are chosen to lie along boundaries of the input domain or output range classes, data structures, procedure parameters, etc. Choices often include maximum, minimum and trivial values or parameters. This technique is often called stress testing.

See Testing, boundary value.

Source: NBS

Branch—An instruction which causes program execution to jump to a new point in the program sequence, rather than execute the next instruction. Contrasts with condition coverage, multiple condition coverage, path coverage and statement coverage.

See Decision coverage.

Branch coverage—A test coverage criterion that requires that for each decision point, each possible branch is executed at least once. Synonymous with decision coverage. Contrasts with condition coverage, multiple condition coverage, path coverage and statement coverage.

Buffer—A device or storage area (memory) used to store data temporarily to compensate for differences in rates of data flow, time of occurrence of events or amounts of data that can be handled by the devices or processes involved in the transfer or use of the data

Bug—A fault in a program that causes the program to perform in an unintended or unanticipated manner

See Anomaly, Defect, Error, Exception and Fault.

Byte—A sequence of adjacent bits, often an octet, operated on as a unit

C

C—A general-purpose high-level programming language that was created for use in the development of computer operating systems software. It strives to combine the power of assembly language with the ease of a high-level language.

C++—An object-oriented high-level programming language

CD-ROM—Compact disc, read-only memory

Central processing unit (CPU)—Computer hardware that houses the electronic circuits that control/direct all operations of the computer system.

Change control—The processes, authorities and procedures to be used for all changes that are made to the computerized system and/or the system data. Change control is a vital subset of the quality assurance (QA) program in an enterprise and should be clearly described in the enterprise standard operating procedures (SOPs).

See Configuration control.

Client-server—A term used in a broad sense to describe the relationship between the receiver and the provider of a service. Generally, the client-server describes a networked system where front-end applications, like the client, make service requests to another networked system. Client-server relationships are defined primarily by software. In a local area network (LAN), the workstation is the client and the file server is the server. However, client-server systems are inherently more complex than file server systems. Two disparate programs must work in tandem, and there are many more decisions to make about separating data and processing between the client workstations and the database server. The database server encapsulates database files and indexes, restricts access, enforces security and provides applications with a consistent interface to data via a data dictionary.

COBOL—Common Business Oriented Language (COBOL) is a high-level programming language intended for use in the solution of problems in business data processing.

Code audit—An independent review of source code by a person, team or tool to verify compliance with software design documentation and programming standards. Correctness and efficiency may also be evaluated. Contrasts with code inspection, code review and code walkthrough.

Coding— 1. In software engineering, the process of expressing a computer program in a programming language

2. The transforming of logic and data from design specifications (design descriptions) into a programming language

Coding standards—Written procedures describing coding (programming) style conventions that specify rules governing the use of individual constructs provided by the programming language and naming, formatting, and documentation requirements, which prevent programming errors, control complexity and promote understandability of the source code. Synonymous with development standards and programming standards.

Comment—1. In programming languages, a language construct that allows explanatory text to be inserted into a program and that does not have any effect on the execution of the program

2. Information embedded within a computer program, job control statements or a set of data that provides clarification to human readers but does not affect machine interpretation

Source: IEEE

Compact disc – read-only memory—A compact disk used for the permanent storage of text, graphic or sound information. Digital data is represented very compactly by tiny holes that can be read by lasers attached to high resolution sensors. Capable of storing up to 680 MB of data, equivalent to 250,000 pages of text, or 20,000 medium resolution images. This storage medium is often used for archival purposes. Synonymous with optical disk and write-once read-many times disk.

Compilation—Translating a program expressed in a problem-oriented language or a procedure-oriented language into object code. Contrasts with assembling and interpret.

Compiler—1. A computer program that translates programs expressed in a high-level language into their machine-language equivalents

2. The compiler takes the finished source-code listing as input and outputs the machine-code instructions that the computer must have to execute the program.

See Assembler and Interpreter

Computer—1. A functional unit that can perform substantial computations, including numerous arithmetic operations, or logic operations, without human intervention during a run

2. A functional programmable unit that consists of one or more associated processing units and peripheral equipment, is controlled by internally stored programs, and can perform substantial computations, including numerous arithmetic operations, or logic operations, without human intervention

Computer instruction set—A complete set of the operators of the instructions of a computer together with a description of the types of meanings that can be attributed to their operands. Synonymous with machine instruction set.

Computer language—A language designed to enable humans to communicate with computers

See Programming language.

Computer science—The branch of science and technology that is concerned with methods and techniques relating to data processing performed by automatic means

Computer system— A functional unit, consisting of one or more computers, associated peripheral input and output devices, and associated software, that uses common storage for all or part of a program and for all or part of the data necessary for the execution of the program; executes user-written or user-designated programs; performs user-designated data manipulation, including arithmetic operations and logic operations; and can execute programs that modify themselves during their execution. A computer system may be a stand-alone unit or may consist of several interconnected units.

See Computer.

Configuration management—The control of changes to a set of configuration items over a system life cycle

Consistency— The degree of uniformity, standardization and freedom from contradiction among the documents or parts of a system or component

See Traceability.

Consistency checker—A software tool used to test requirements in design specifications for both consistency and completeness

Constant—A value that does not change during processing. Contrasts with variable.

Control flow diagram—A diagram that depicts the set of all possible sequences in which operations may be performed during the execution of a system or program. Types include box diagram, flowchart, input-process-output chart and state diagram. Contrasts with data flow diagram.

COTS—Configurable, off-the-shelf software

Coverage analysis—Determining and assessing measures associated with the invocation of program structural elements to determine the adequacy of a test run. Coverage analysis is useful when attempting to execute each statement, branch, path or iterative structure in a program. Tools that capture this data and provide reports summarizing relevant information have this feature.

See Testing, branch; Testing, path and Testing, statement.

CPU—See Central processing unit.

Crash—The sudden and complete failure of a computer system or component

Critical control point—(QA) A function or an area in a manufacturing process or procedure, the failure of which, or loss of control over, may have an adverse effect on the quality of the finished product and may result in an unacceptable health risk

Critical design review— A review conducted to verify that the detailed design of one or more configuration items satisfy specified requirements; to establish the compatibility among the configuration items and other items of equipment, facilities, software, and personnel; to assess risk areas for each configuration item; and, as applicable, to assess the results of producibility analyses, review preliminary hardware product specifications, evaluate preliminary test planning and evaluate the adequacy of preliminary operation and support documents

See System design review.

D

Data—Representations of facts, concepts or instructions in a manner suitable for communication, interpretation or processing by humans or by automated means. In the simplest terms, data are pieces of information.

Data analysis—Obtaining an understanding of data by considering samples, measurement and visualization. Data analysis can be particularly useful when a data set is first received, before one builds the first model. It is also crucial in understanding experiments and debugging problems with the system.

Data exception—An exception that occurs when a program attempts to use or access data incorrectly

Data flow analysis—A software verification and validation (V&V) task to ensure that the input and output data and their formats are properly defined, and that the data flows are correct

Data flow diagram—A diagram that depicts data sources, data sinks, data storage, processes performed on data as nodes and logical flow of data as links between the nodes

Data integrity—The degree to which a collection of data is complete, consistent and accurate

Data set—A collection of related records

Data validation—1. A process used to determine if data are inaccurate, incomplete or unreasonable. The process may include format checks, completeness checks, check key tests, reasonableness checks and limit checks.

2. The checking of data for correctness or compliance with applicable standards, rules and conventions.

Database—A collection of data, often with controlled redundancy, organized according to a schema to serve one or more applications. The data are stored so that they can be used by different programs without concern for the data structure or organization. A common approach is used to add new data and to modify and retrieve existing data. See Archival database.

Database analysis—A software verification and validation (V&V) task to ensure that the database structure and access methods are compatible with the logical design

Database security—The degree to which a database is protected from exposure to accidental or malicious alteration or destruction

Debugging—Determining the exact nature and location of a program error and fixing the error

Decision coverage—A test coverage criterion requiring enough test cases so that each decision has a true and false result at least once and each statement is executed at least once. Synonymous with branch coverage. Contrasts with condition coverage, multiple condition coverage, path coverage, statement coverage.

Default value—A standard setting or state to be taken by the program if no alternate setting or state is initiated by the system or the user. A value assigned automatically if one is not given by the user.

Defect—See Bug, Error, Exception and Fault.

Design—The process of defining the architecture, components, interfaces and other characteristics of a system or component

See Architectural design, Preliminary design and detailed design.

Design phase—The period of time in the software life cycle during which the designs for architecture, software components, interfaces and data are created, documented and verified to satisfy requirements

Developer—A person or group that designs. and/or builds, and/or documents and/or configures the hardware and/or software of computerized systems

Development methodology—A systematic approach to software creation that defines development phases and specifies the activities, products, verification procedures and completion criteria for each phase

See Incremental development, Rapid prototyping, Spiral model and Waterfall model.

Diagnostic—Pertaining to the detection and isolation of faults or failures, e.g. a diagnostic message and a diagnostic manual

Disk—Circular rotating magnetic storage hardware. Disks can be hard (fixed) or flexible (removable), and different sizes

Documentation—The aids provided for the understanding of the structure and intended uses of an information system or its components, such as flowcharts, textual material and user manuals

Documentation, software—Technical data or information, including computer listings and printouts, in human readable form, that describe or specify the design or details, explain the capabilities, or provide operating instructions for using the software to obtain desired results from a software system

See Specification; Specification, requirements; Specification, design; Software design description; Test plan, Test report and User's guide.

Driver—A program that links a peripheral device or internal function to the operating system and provides for activation of all device functions. Contrasts with test driver.

E

Embedded software—Software that is part of a larger system and performs some of the requirements of that system, e.g., software used in an aircraft or rapid transit system. Such software does not provide an interface with the user. See Firmware.

Encapsulation—The technique used by layered protocols in which a lower-layer protocol accepts a message from a higher-layer protocol and places it in the data portion of a frame in the lower layer. In software development, it is a technique that isolates a system function or a set of data, and the operations on those data, within a module, and provides precise specifications for the module. See Abstraction, Information hiding and Software engineering.

End user— 1. A person, device, program or computer system that uses an information system for the purpose of data processing in information exchange

2. A person whose occupation requires the use of an information system, but does not require any knowledge of computers or computer programming

See User.

Entity relationship diagram—A diagram that depicts a set of real-world entities and the logical relationships among them

Environment—1. Everything that supports a system or the performance of a function

2. The conditions that affect the performance of a system or function

Error— A deviation between a computed, observed, or measured value or condition and the true, specified, or theoretically correct value or condition

See Anomaly, Bug, Defect, Exception and Fault.

Error detection—Techniques used to identify errors in data transfers

Event table—A table that lists events and the corresponding specified effect(s) of or reaction(s) to each event

Exception—An event that causes suspension of normal program execution. Types include addressing exception, data exception, operation exception, overflow exception, protection exception and underflow exception.

F

Fail-safe— A system or component that automatically places itself in a safe operational mode in the event of a failure

Source: IEEE

Failure—The inability of a system or component to perform its required functions within specified performance requirements

Source: IEEE

See Bug, Crash, Exception and Fault.

Failure analysis—Determining the exact nature and location of a program error to fix the error, to identify and fix other similar errors, and to initiate corrective action to prevent future occurrences of this type of error. Contrasts with debugging.

Feasibility study—Analysis of the known or anticipated need for a product, system or component to assess the degree to which the requirements, designs or plans can be implemented

Field— 1. On a data medium or in storage, a specified area used for a particular class of data, e.g., a group of character positions used to enter or display wage rates on a screen

2. Defined logical data that is part of a record

3. The elementary unit of a record that may contain a data item, a data aggregate, a pointer or a link

4. A discrete location in a database that contains a unique piece of information. A field is a component of a record. A record is a component of a database.

File— 1. A set of related records treated as a unit, e.g., in stock control, a file can consist of a set of invoices

2. The largest unit of storage structure that consists of a named collection of all occurrences in a database of records of a particular record type

File Transfer Protocol (FTP)—A protocol used to transfer files over a Transmission Control Protocol/Internet Protocol (TCP/IP) network (Internet, UNIX, etc.).

Firmware— The combination of a hardware device, e.g., an IC, and computer instructions and data that reside as read only software on that device. Such software cannot be modified by the computer during processing.

See Embedded software.

Flat file—A data file that does not physically interconnect with or point to other files. Any relationship between two flat files is logical, e.g., matching account numbers.

Flowchart or flow diagram—1. Graphical representation in which symbols are used to represent such things as operations, data, flow direction and equipment, for the definition, analysis or solution of a problem

2. A control flow diagram in which suitably annotated geometrical figures are used to represent operations, data or equipment, and arrows are used to indicate the sequential flow from one to another. Synonymous with flow diagram.

See Block diagram, Box diagram, Bubble chart, Graph, Input-Process-output chart and Structure chart.

FORTRAN—An acronym for FORmula TRANslator, the first widely used high-level programming language. Intended primarily for use in solving technical problems in mathematics, engineering and science

Function—1. A mathematical entity whose value, namely, the value of the dependent variable, depends in a specified manner on the values of one or more independent variables, with not more than one value of the dependent variable corresponding to each permissible combination of values from the respective ranges of the independent variables

2. A specific purpose of an entity, or its characteristic action

3. In data communication, a machine action, such as carriage return or line feed

Functional analysis—Verifies that each safety-critical software requirement is covered and that an appropriate criticality level is assigned to each software element

Functional design—1. The process of defining the working relationships among the components of a system. See Architectural design.

2. The result of the process in definition 1

Functional requirement—A requirement that specifies a function that a system or system component must be able to perform

G

GB—Gigabyte

Gigabyte—Approximately one-billion bytes; precisely 2^{30} or 1,073,741,824 bytes

See Kilobyte and Megabyte.

Graph— A diagram or other representation consisting of a finite set of nodes and internode connections called edges or arcs. Contrasts with blueprint.

See Block diagram, Box diagram, Bubble chart, Call graph, Cause-effect graph, Control flow diagram, Data flow diagram, Directed graph, Flowchart, Input-process-output chart, Structure chart and Transaction flowgraph.

Graphic software specifications—Documents, such as charts, diagrams, graphs that depict program structure, states of data, control, transaction flow, HIPO, and cause-effect relationships; and tables, including truth, decision, event, state-transition, module interface, exception conditions/responses, necessary to establish design integrity

H

Hard disk drive—Hardware used to read from or write to a hard disk

See Disk and Disk drive.

Hardware—Physical equipment, as opposed to programs, procedures, rules and associated documentation. Contrasts with software.

Hexadecimal—The base-16 number system. Digits are 0, 1, 2, 3, 4, 5, 6, 7, 8, 9, A, B, C, D, E, & F. This is a convenient form in which to examine binary data, because it collects four binary digits per hexadecimal digit, e.g., decimal 15 is 1111 in binary and F in hexadecimal.

High-level language—A programming language that requires little knowledge of the target computer, can be translated into several different machine languages, allows symbolic naming of operations and addresses, provides features designed to facilitate expression of data structures and program logic, and usually results in several machine instructions for each program statement. Examples are PL/1, COBOL, BASIC, FORTRAN, Ada, Pascal and C. Contrasts with assembly language.

I

I/O—Input/output

IEEE—Pronounced I-triple-E; IEEE is an organization composed of engineers, scientists and students. Best known for developing standards for the computer and electronics.

Implementation—The process of translating a design into hardware components, software components or both

See Coding.

Implementation phase—The period of time in the software life cycle during which a software product is created from design documentation and debugged

Incremental development—A software development technique in which requirements definition, design, implementation and testing occur in an overlapping, iterative (rather than sequential) manner, resulting in incremental completion of the overall software product. Contrasts with rapid prototyping, spiral model and waterfall model.

Incremental integration—A structured reformation of the program, module by module or function by function, with an integration test being performed following each addition. Methods include top-down, breadth-first, depth-first and bottom-up.

Industry standard—Procedures and criteria recognized as acceptable practices by peer professional, credentialing or accrediting organizations

Information hiding—The practice of hiding the details of a function or structure, making them inaccessible to other parts of the program

See Abstraction, Encapsulation and Software engineering.

Input-processing-output—A structured software design technique; identification of the steps involved in each process to be performed and identifying the inputs to and outputs from each step. A refinement called hierarchical input-process-output identifies the steps, inputs and outputs at both general and detailed levels of detail.

Input/output—Each microprocessor and each computer needs a way to communicate with the outside world to get the data needed for its programs and to communicate the results of its data manipulations. This is accomplished through I/0 ports and devices.

Installation—The phase in the system life cycle that includes assembly and testing of the hardware and software of a computerized system. Installation includes installing a new computer system, new software or hardware, or otherwise modifying the current system.

Installation and checkout phase—The period of time in the software life cycle during which a software product is integrated into its operational environment and tested in this environment to ensure that it performs as required

Institute of Electrical and Electronic Engineers (IEEE)—An organization involved in the generation and promulgation of standards. IEEE standards represent the formalization of current norms of professional practice through the process of obtaining the consensus of concerned, practicing professionals in the given field

Instruction—1. A program statement that causes a computer to perform a particular operation or set of operations

2. In a programming language, a meaningful expression that specifies one operation and identifies its operands, if any

Instruction set—1. The complete set of instructions recognized by a given computer or provided by a given programming language

2. The set of the instructions of a computer, of a programming language or of the programming languages in a programming system

See Computer instruction set.

Interactive—Pertaining to a system or mode of operation in which each user entry causes a response from or action by the system. Contrasts with batch.

See Conversational, Online and Real time.

Interface—1. A shared boundary between two functional units, defined by functional characteristics, common physical interconnection characteristics, signal characteristics and other characteristics, as appropriate. The concept involves the specification of the connection of two devices having different functions.

2. A point of communication between two or more processes, persons, or other physical entities.

3. A peripheral device that permits two or more devices to communicate.

International Organization for Standardization (ISO)—An organization that sets international standards. It deals with all fields except electrical and electronics, which are governed by the International Electrotechnical Commission (IEC). Synonymous with International Standards Organization.

Interpret—To translate and execute each statement or construct of a computer program before translating and executing the next. Contrasts with assemble and compile.

Invalid inputs—1. Test data that lie outside the domain of the function the program represents

2. These are not only inputs outside the valid range for data to be input, i.e. when the specified input range is 50 to 100, but also unexpected inputs, especially when these unexpected inputs may easily occur, e.g., the entry of alpha characters or special keyboard characters when only numeric data is valid, or the input of abnormal command sequences to a program.

ISO—International Organization for Standardization

J

Job—A user-defined unit of work that is to be accomplished by a computer. For example, the compilation, loading, and execution of a computer program.

See Job control language.

K

KB—Kilobyte

Kilobyte—Approximately one-thousand bytes. This symbol is used to describe the size of computer memory or disk storage space. Because computers use a binary number system, a kilobyte is precisely 2^{10} or 1024 bytes.

L

Life cycle methodology—The use of any one of several structured methods to plan, design, implement, test and operate a system from its conception to the termination of its use.

See Waterfall model.

Low-level language—See Assembly language. The advantage of assembly language is that it provides bit-level control of the processor, allowing tuning of the program for optimal speed and performance. For time-critical operations, assembly language may be necessary to generate code that executes fast enough for the required operations. The disadvantage of assembly language is the high-level of complexity and detail required in the programming. This makes the source code harder to understand, thus increasing the chance of introducing errors during program development and maintenance.

M

Machine code—Computer instructions and definitions expressed in a form [binary code] that can be recognized by the CPU of a computer. All source code, regardless of the language in which it was programmed, is eventually converted to machine code.

Main memory—A nonmoving storage device utilizing one of a number of types of electronic circuitry to store information

Main program—A software component that is called by the operating system of a computer and that usually calls other software components

See Routine and Subprogram.

Mainframe—A large high-speed computer, especially one supporting numerous workstations or peripherals

Maintainability—The ease with which a software system or component can be modified to correct faults, improve performance or other attributes, or adapt to a changed environment. Synonymous with modifiability.

Maintenance—(QA) Activities, such as adjusting, cleaning, modifying, and overhauling equipment to assure performance in accordance with requirements. Maintenance to a software system includes correcting software errors, adapting software to a new environment or making enhancements to software.

See Adaptive maintenance, corrective maintenance and Perfective maintenance.

MB—Megabyte

Measurement—The process of determining the value of some quantity in terms of a standard unit

Measure (IEEE) —A quantitative assessment of the degree to which a software product or process possesses a given attribute

Megabit—Approximately one million bits. Precisely 1024 K bits, 2^{20} bits, or 1,048,576 bits.

Megabyte—Approximately one million bytes. Precisely 1024 K Bytes, 2^{20} bytes, or 1,048,576 bytes.

See Kilobyte.

Memory—Any device or recording medium into which binary data can be stored and held, and from which the entire original data can be retrieved. The two types of memory are main, e.g., ROM and RAM, and auxiliary, e.g., tape and disk. —Any device or recording medium into which binary data can be stored and held, and from which the entire original data can be retrieved. The two types of memory are main, e.g., ROM and RAM, and auxiliary, e.g., tape and disk.

See Storage device.

Metric—A quantifiable entity that allows the measurement of the achievement of a process goal

Scope Notes: Metrics should be SMART—specific, measurable, actionable, relevant and timely. Complete metric guidance defines the unit used, measurement frequency, ideal target value (if appropriate) and also the procedure to carry out the measurement and the procedure for the interpretation of the assessment.

Mnemonic—A symbol chosen to assist human memory and understanding, e.g., an abbreviation such as MPY for multiply

Modeling—Construction of programs used to model the effects of a postulated environment for investigating the dimensions of a problem for the effects of algorithmic processes on responsive targets

Modular software—Software composed of discrete parts

Modularity—The degree to which a system or computer program is composed of discrete components so that a change to one component has minimal impact on other components

Module—1. In programming languages, a self-contained subdivision of a program that may be separately compiled

2. A discrete set of instructions, usually processed as a unit by an assembler, a compiler, a linkage editor, or similar routine or subroutine

3. A packaged functional hardware unit suitable for use with other components

See Unit.

Multiprocessing—A mode of operation in which two or more processes [programs] are executed concurrently [simultaneously] by separate CPUs that have access to a common main memory. Contrasts with multiprogramming.

See Multitasking and time sharing.

N

National Institute for Standards and Technology (NIST)—Develops tests, test methods, reference data, proof-of concept implementations, and technical analyses to advance the development and productive use of information technology.

Scope Notes: NIST is a US government entity that creates mandatory standards that are followed by federal agencies and those doing business with them.

Network—A system of interconnected computers and the communication equipment used to connect them.

Null—A value whose definition is to be supplied within the context of a specific operating system. This value is a representation of the set of no numbers or no value for the operating system in use.

Null data—Data for which space is allocated, but for which no value currently exists

Null string—A string containing no entries. Note that a null string has length zero.

O

Object oriented design—A software development technique in which a system or component is expressed in terms of objects and connections between those objects

Object oriented language—A programming language that allows the user to express a program in terms of objects and messages between those objects. Examples include C++, Smalltalk and LOGO.

Object oriented programming—A technology for writing programs that are made up of self-sufficient modules that contain all the information needed to manipulate a given data structure. The modules are created in class hierarchies, so that the code or methods of a class can be passed to other modules. New object modules can be easily created by inheriting the characteristics of existing classes.

See Object and Object-oriented design.

Octal—The base-8 number system. Digits are 0, 1, 2, 3, 4, 5, 6 and 7.

OOP—Object-oriented programming.

Operating system—Software that controls the execution of programs and provides services, such as resource allocation, scheduling, input/output control and data management. Usually, operating systems are predominantly software, but partial or complete hardware implementations are possible.

Operation and maintenance phase—The period of time in the software life cycle during which a software product is employed in its operational environment, monitored for satisfactory performance, and modified as necessary to correct problems or to respond to changing requirements

Oracle—A relational-database programming system incorporating the SQL programming language. A registered trademark of the Oracle Corp.

Overflow—In a calculator, the state in which the calculator is unable to accept or process the number of digits in the entry or in the result

Source: ISO

See Arithmetic overflow.

Overflow exception—An exception that occurs when the result of an arithmetic operation exceeds the size of the storage location that is designated to receive it

Source: IEEE

P

Parameter—A constant, variable or expression that is used to pass values between software modules. Synonymous with argument.

Pascal—A high-level programming language designed to encourage structured programming practices

Patch—Fixes to software programming errors and vulnerabilities

Path—A sequence of instructions that may be performed in the execution of a computer program

Path analysis—Analysis of a computer program (i.e., source code) to identify all possible paths through the program, to detect incomplete paths or discover portions of the program that are not on any path

Peripheral device—Equipment that is directly connected to a computer. A peripheral device can be used to input data, e.g., keypad, bar code reader, transducer and laboratory test equipment, or to output data, e.g., printer, disk drive, video system, tape drive, valve controller and motor controller. Synonymous with peripheral equipment.

Personal computer (PC)—Synonymous with microcomputer, a computer that is functionally similar to large computers, but serves only one user

Platform—The hardware and software that must be present and functioning for an application program to run (perform) as intended. A platform includes, but is not limited to, the operating system or executive software, communication software, microprocessor, network, input/output hardware, any generic software libraries, database management, user interface software, and the like.

Polynomial—Mathematical expression of more than two algebraic terms, especially the sum of several terms that contain different powers of the same variable(s).

Preliminary design—1. The process of analyzing design alternatives and defining the architecture, components, interfaces, and timing and sizing estimates for a system or component

See Detailed design.

2. The result of the process in definition 1

Program (IT)—1. A sequence of instructions suitable for processing. Processing may include the use of an assembler, compiler, interpreter or another translator to prepare the program for execution. The instructions may include statements and necessary declarations.

2. (ISO) To design, write, and test programs.

3. (ANSI) In programming languages, a set of one or more interrelated modules capable of being executed.

4. Loosely, a routine.

5. Loosely, to write a routine.

Programmable read-only memory (PROM)—A chip that can be programmed using a PROM programming device. It can be programmed only once. It cannot be erased and reprogrammed. Each of its bit locations is a fusible link. An unprogrammed PROM has all links closed, establishing a known state of each bit. Programming the chip consists of sending an electrical current of a specified size through each link that is to be changed to the alternate state. This causes the fuse to blow, opening that link.

Programming language—A language used to express computer programs

Source: IEEE

See Computer language, High-level language and Low-level language.

Project plan—A management document describing the approach taken for a project. The plan typically describes work to be done, resources required, methods to be used, the configuration management and quality assurance procedures to be followed, the schedules to be met, the project organization, etc. Project in this context is a generic term. Some projects may also need integration plans, security plans, test plans, quality assurance plans, etc.

Source: NIST

See Documentation plan, Software development plan, Test plan and Software engineering.

Prototyping—The process of quickly putting together a working model (a prototype) to test various aspects of a design, illustrate ideas or features, and gather early user feedback. Prototyping uses programmed simulation techniques to represent a model of the final system to the user for advisement and critique. The emphasis is on end-user screens and reports. Internal controls are not a priority item since this is only a model.

Pseudocode—A combination of programming language and natural language used to express a software design. If used, it is usually the last document produced prior to writing the source code.

Q

QA—Quality assurance

QC—Quality control

Quality assurance (QA)—A planned and systematic pattern of all actions necessary to provide adequate confidence that an item or product conforms to established technical requirements. (ISO/IEC 24765)

Quality assurance, software—1. A planned and systematic pattern of all actions necessary to provide adequate confidence that an item or product conforms to established technical requirements

2. A set of activities designed to evaluate the process by which products are developed or manufactured

Quality control—The operational techniques and procedures used to achieve quality requirements

R

RAM—Random access memory

Random access memory (RAM)—TA type of primary computer memory. RAM is volatile, and data is lost with power loss.

Rapid prototyping—A structured software requirements discovery technique that emphasizes generating prototypes early in the development process to permit early feedback and analysis in support of the development process. Contrasts with incremental development, spiral model and waterfall model.

See Prototyping.

Read-only memory (ROM)—A type of primary computer memory. ROM is nonvolatile, and data stored there survives power loss.

Real-time processing—A fast-response (immediate response) online system that obtains data from an activity or a physical process, performs computations and returns a response rapidly enough to affect (i.e., control) the outcome of the activity or process, e.g., a process control application. Contrasts with batch processing.

Record—A collection of related information that is treated as a unit. Separate fields within the record are used for processing of the information.

Regression analysis and testing—A software verification and validation (V&V) task to determine the extent of V&V analysis and testing that must be repeated when changes are made to any previously examined software products

Source: IEEE

See Testing, regression.

Relational database—Database organization method that links files together as required. Relationships between files are created by comparing data, such as account numbers and names. A relational system can take any two or more files and generate a new file from the records that meet the matching criteria. Routine queries often involve more than one data file, e.g., a customer file and an order file can be linked to ask a question that relates to information in both files, such as the names of the customers that purchased a particular product. Contrasts with network database and flat file.

Release—The formal notification and distribution of an approved version

See Version.

Requirement—1. A condition or capability needed by a user to solve a problem or achieve an objective

2. A condition or capability that must be met or possessed by a system or system component to satisfy a contract, standard, specification or other formally imposed documents

3. A documented representation of a condition or capability as in definition 1 or 2

See Design requirement, Functional requirement, Implementation requirement, Interface requirement, Performance requirement and Physical requirement.

Requirements analysis—1. The process of studying user needs to arrive at a definition of a system, hardware or software requirements

2. The process of studying and refining system, hardware or software requirements

Source: IEEE

See Prototyping and Software engineering.

Requirements phase—The period of time in the software life cycle during which the requirements, such as functional and performance capabilities for a software product, are defined and documented

Source: IEEE

Review—A process or meeting during which a work product or set of work products is presented to project personnel, managers, users, customers or other interested parties for comment or approval. Types include code review, design review, formal qualification review, requirements review and test readiness review. Contrasts with audit and inspection.

Source: IEEE

See Static analysis.

Risk—The combination of the likelihood of an event and its impact

Risk assessment—A process used to identify and evaluate risk and its potential effects

Scope Notes: Risk assessments are used to identify those items or areas that present the highest risk, vulnerability or exposure to the enterprise for inclusion in the IS annual audit plan.

Risk assessments are also used to manage the project delivery and project benefit risk.

Robustness—The degree to which a software system or component can function correctly in the presence of invalid inputs or stressful environmental conditions

See Software reliability.

ROM—See Read-only memory.

Routine—A subprogram that is called by other programs and subprograms

Scope Notes: This term is defined differently in various programming languages.

Source: IEEE

See Module.

S

Server—A high-speed computer in a network that is shared by multiple users. It holds the programs and data that are shared by all users.

Software—Programs, procedures, rules and any associated documentation pertaining to the operation of a system. Contrasts with hardware. See Application software, Operating system, System software and Utility software.

Software development plan—The project plan for the development of a software product. Contrasts with software development process and software life cycle.

Software development process—The process by which user needs are translated into a software product. The process involves translating user needs into software requirements, transforming the software requirements into design, implementing the design in code, testing the code and sometimes installing and checking out the software for operational activities. Note that these activities may overlap or be performed iteratively.

See Incremental development, Rapid prototyping, Spiral model and Waterfall model.

Software documentation—Technical data or information, including computer listings and printouts, in human-readable form, that describe or specify the design or details, explain the capabilities, or provide operating instructions for using the software to obtain desired results from a software system. Types of software documentation include:

- Project planning documents, i.e., software development plans and software verification and validation (V&V) plans
- Software requirements and design specifications
- Test documentation
- Customer-deliverable documentation
- Program source code
- Representation of software solutions implemented in firmware
- Reports, e.g., review, audit and project status
- Data, i.e., defect detection and test
- Contrasts with software item.
- See: specification; specification, requirements; specification, design; software design description; test plan, test report, user's guide.

Software engineering—The application of a systematic, disciplined, quantifiable approach to the development, operation and maintenance of software, i.e., the application of engineering to software

See Project plan, Requirements analysis, Architectural design, Structured design, System safety, Testing and Configuration management.

Software engineering environment—The hardware, software and firmware used to perform a software engineering effort. Typical elements include computer equipment, compilers, assemblers, operating systems, debuggers, simulators, emulators, test tools, documentation tools and database management systems.

Software life cycle— Period of time, beginning when a software product is conceived and ending when the product is no longer available for use. The software life cycle is typically broken into phases, denoting activities, such as requirements, design, programming, testing, installation, operation and maintenance. Contrasts with software development process.

See Waterfall model.

Software reliability—1. The probability that software will not cause the failure of a system for a specified time under specified conditions. The probability is a function of the inputs to and use of the system in the software. The inputs to the system determine whether existing faults, if any, are encountered.

2. The ability of a program to perform its required functions accurately and reproducibly under stated conditions for a specified period of time

Software review—An evaluation of software elements to ascertain discrepancies from planned results and to recommend improvement. This evaluation follows a formal process. Synonymous with software audit.

See Code audit, Code inspection, Code review, Code walkthrough, Design review, Specification analysis and Static analysis.

SOPs—Standard operating procedures

Source code—Computer instructions and data definitions expressed in a form that is suitable for input to an assembler, compiler or other translator

Source program—A computer program that must be compiled, assembled or otherwise translated to be executed by a computer. Contrasts with object program.

See Source code.

Spaghetti code—Program source code written without a coherent structure. Implies the excessive use of GOTO instructions. Contrasts with structured programming.

Specification tree—A diagram that depicts all the specifications for a given system and shows their relationship to one another

Source: IEEE

Specification, requirements—A specification that documents the requirements of a system or system component. It typically includes functional requirements, performance requirements, interface requirements, design requirements (attributes and constraints), development (coding) standards, etc. Contrasts with requirement.

Source: NIST

Spiral model—A model of the software development process in which the constituent activities, typically requirements analysis, preliminary and detailed design, coding, integration and testing, are performed iteratively until the software is complete. Synonymous with evolutionary model. Contrasts with incremental development, rapid prototyping and waterfall model.

SQL—The ISO standard query language used by application programmers and end users to access relational databases. Variations of this popular language are often available for data storage systems that are not strictly relational.

Standard operating procedures—Written procedures that prescribe and describe the steps to be taken in normal and defined conditions and that are necessary to ensure control of production and processes

State—A condition or mode of existence in which a system, component or simulation may be, e.g., the preflight state of an aircraft navigation program or the input state of a given channel

State diagram—A diagram that depicts the states that a system or component can assume and shows the events or circumstances that cause or result from a change from one state to another

Synonymous with state graph.

See State-transition table.

Storage device—A unit into which data or programs can be placed, retained and retrieved

See Memory.

String—1. A sequence of characters

2. A linear sequence of entities, such as characters or physical elements

Structured design—Any disciplined approach to software design that adheres to specified rules based on principles, such as modularity, top-down design and stepwise refinement of data; system structure and processing steps

See Data structure centered design, Input-processing-output, Modular decomposition, Object-oriented design, Rapid prototyping, Stepwise refinement, Structured programming, Transaction analysis, Transform analysis, Graphical software specification/design documents, Modular software and Software engineering.

Structured programming—Any software development technique that includes structured design and results in the development of structured programs

See Structured design.

Structured Query Language (SQL)—A language used to interrogate and process data in a relational database. Originally developed for IBM mainframes, many implementations have been created for mini- and microcomputer database applications. SQL commands can be used to interactively work with a database or can be embedded with a programming language to interface with a database.

Subprogram—A separately compilable, executable component of a computer program. Note that this term is defined differently in various programming languages.

See Coroutine, Main program, Routine and Subroutine.

Subroutine—A routine that returns control to the program or subprogram that called it. Note that this term is defined differently in various programming languages.

See Module.

Subroutine trace—A record of all or selected subroutines or function calls performed during the execution of a computer program and, optionally, the values of parameters passed to and returned by each subroutine or function

Support software—Software that aids in the development and maintenance of other software, e.g., compilers, loaders and other utilities

Synchronous—Occurring at regular, timed intervals, i.e., timing dependent

Synchronous transmission—Block-at-a-time data transmission.

Syntax—The structural or grammatical rules that define how symbols in a language are to be combined to form words, phrases, expressions and other allowable constructs

System—1. People, machines and methods organized to accomplish a set of specific functions

2. (DOD) A composite, at any level of complexity, of personnel, procedures, materials, tools, equipment, facilities and software. The elements of this composite entity are used together in the intended operational or support environment to perform a given task or achieve a specific purpose, support or mission requirement.

System analysis—A systematic investigation of a real or planned system to determine the functions of the system and how they relate to each other and to any other system

See Requirements phase.

System design—A process of defining the hardware and software architecture, components, modules, interfaces and data for a system to satisfy specified requirements

See Design phase, Architectural design and Functional design.

System design review—A review conducted to evaluate the manner in which the requirements for a system have been allocated to configuration items, the system engineering process that produced the allocation, the engineering planning for the next phase of the effort, manufacturing considerations and the planning for production engineering

Source: IEEE

See Design review.

System documentation—The collection of documents that describe the requirements, capabilities, limitations, design, operation and maintenance of an information processing system

Source: ISO

See Specification, Test documentation and User's guide.

System integration—The progressive linking and testing of system components into a complete system—The progressive linking and testing of system components into a complete system

Source: ISO

See Incremental integration.

System life cycle—The course of developmental changes through which a system passes from its conception to the termination of its use, e.g., the phases and activities associated with the analysis, acquisition, design, development, test, integration, operation, maintenance and modification of a system

See Software life cycle.

System software—1. Application-independent software that supports the running of application software

2. Software designed to facilitate the operation and maintenance of a computer system and its associated programs, e.g., operating systems, assemblers and utilities. Contrasts with application software.

See Support software.

T

TB—Terabyte

Terabyte—Approximately one-trillion bytes; precisely 2^{40} or 1,099,511,627,776 bytes

See Kilobyte, Megabyte and Gigabyte.

Terminal—A device, usually equipped with a CRT display and keyboard, used to send and receive information to and from a computer via a communication channel

Test—An activity in which a system or component is executed under specified conditions, the results are observed or recorded, and an evaluation is made of some aspect of the system or component

Test case—Documentation specifying inputs, predicted results and a set of execution conditions for a test item

Test case generator—A software tool that accepts as input source code, test criteria, specifications or data structure definitions; uses these inputs to generate test input data; and, sometimes, determines expected results

Synonymous with test data generator and test generator.

Test design—Documentation specifying the details of the test approach for a software feature or combination of software features and identifying the associated tests

See Testing functional; Cause effect graphing; Boundary value analysis; Equivalence class partitioning; Error guessing; Testing, structural; Branch analysis; Path analysis; Statement coverage; Condition coverage; Decision coverage and Multiple-condition coverage.

Test documentation—Documentation describing plans for, or results of, the testing of a system or component. Types include test case specification, test incident report, test log, test plan, test procedure and test report.

Test driver—A software module used to invoke a module under test and, often, provide test inputs, control and monitor execution, and report test results

Synonymous with test harness.

Test item—A software item that is the object of testing

Test log—A chronological record of all relevant details about the execution of a test

Test phase—The period of time in the software life cycle in which the components of a software product are evaluated and integrated, and the software product is evaluated to determine whether or not requirements have been satisfied

Test plan—Documentation specifying the scope, approach, resources and schedule of intended testing activities. It identifies test items, the features to be tested, the testing tasks, responsibilities, required resources and any risk requiring contingency planning.

See Test design and Validation protocol.

Test procedure—A formal document developed from a test plan that presents detailed instructions for the setup, operation and evaluation of the results for each defined test

See Test case.

Test readiness review—1. A review conducted to evaluate preliminary test results for one or more configuration items; to verify that the test procedures for each configuration item are complete, comply with test plans and descriptions, and satisfy test requirements; and to verify that a project is prepared to proceed to formal testing of the configuration items.

2. A review, as in definition1, for any hardware or software component

Contrasts with code review, design review, formal qualification review and requirements review.

Test report—A document describing the conduct and results of the testing carried out for a system or system component

Testability—1. The degree to which a system or component facilitates the establishment of test criteria and the performance of tests to determine whether those criteria have been met

2. The degree to which a requirement is stated in terms that permit establishment of test criteria and performance of tests to determine whether those criteria have been met

See Measurable.

Testing—The examination of a sample from a population to estimate characteristics of the population.

Testing, acceptance—Testing conducted to determine whether a system satisfies its acceptance criteria and to enable the customer to determine whether to accept the system. Contrasts with testing, development and testing, operational.

Testing, alpha—Acceptance testing performed by the customer in a controlled environment at the developer's site. The software is used by the customer in a setting approximating the target environment, with the developer observing and recording errors and usage problems. Source: Pressman.

Testing, beta—1. Acceptance testing performed by the customer in a live application of the software, at one or more end-user sites, in an environment not controlled by the developer. Source: Pressman

2. For medical device software, such use may require an Investigational device exemption [IDE] or Institutional Review Board [IRB] approval.

Testing, boundary value—A testing technique using input values at, just below and just above the defined limits of an input domain; and with input values causing outputs to be at, just below and just above the defined limits of an output domain

See Boundary value analysis and Testing, stress.

Testing, branch—Testing technique to satisfy coverage criteria that require that for each decision point, each possible branch (outcome) is executed at least once. Contrasts with testing, path and testing, statement.

See Branch coverage.

Testing, compatibility—The process of determining the ability of two or more systems to exchange information. In a situation where the developed software replaces an already working program, an investigation should be conducted to assess possible comparability problems between the new software and other programs or systems.

See Different software system analysis; testing, integration and testing, interface.

Testing, design based functional—The application of test data derived through functional analysis that is extended to include design functions and requirement functions

Source: NBS

See Testing, functional.

Testing, development—Testing conducted during the development of a system or component, usually in the development environment, by the developer. Contrasts with testing, acceptance and testing, operational.

Testing, functional—1. Testing that ignores the internal mechanism or structure of a system or component and focuses on the outputs generated in response to selected inputs and execution conditions

2. Testing conducted to evaluate the compliance of a system or component with specified functional requirements and corresponding predicted results

Synonymous with black-box testing and input/output driven testing. Contrasts with testing, structural.

Testing, integration—An orderly progression of testing in which software elements, hardware elements or both are combined and tested to evaluate their interactions, until the entire system has been integrated

Testing, interface—Testing to evaluate whether systems or components pass data and control correctly to one another. Contrasts with testing, unit and testing, system.

See Testing, integration.

Testing, invalid case—A testing technique using erroneous (invalid, abnormal or unexpected) input values or conditions

See Equivalence class partitioning.

Testing, operational—Testing to evaluate a system or component in its operational environment. Contrasts with testing, development and testing, acceptance.

See Testing, system.

Testing, parallel—Testing a new or an altered data processing system with the same source data that is used in another system. The other system is considered as the standard of comparison. Synonymous with parallel run.

Testing, path—Testing to satisfy coverage criterion that each logical path through the program be tested. Often, paths through the program are grouped into a finite set of classes. One path from each class is then tested. Synonymous with path coverage. Contrasts with testing, branch; testing, statement; branch coverage; condition coverage; decision coverage; multiple condition coverage and statement coverage.

Testing, performance—Functional testing to evaluate the compliance of a system or component with specified performance requirements

Testing, regression—Rerunning test cases that a program has previously executed correctly to detect errors spawned by changes or corrections made during software development and maintenance

Testing, special case—A testing technique using input values that seem likely to cause program errors, e.g., 0, 1, NULL and empty string

See Error guessing.

Testing, statement—Testing to satisfy the criterion that each statement in a program be executed at least once during program testing. Synonymous with statement coverage. Contrasts with testing, branch; testing, path; branch coverage; condition coverage; decision coverage; multiple condition coverage and path coverage.

Testing, storage—A determination of whether certain processing conditions use more storage (i.e., memory) than estimated

Testing, stress—Testing to evaluate a system or component at or beyond the limits of its specified requirements. Synonymous with testing, boundary value.

Testing, system—The process of testing an integrated hardware and software system to verify that the system meets its specified requirements. Such testing may be conducted in the development environment and the target environment.

Testing, unit—1. Testing of a module for typographic, syntactic and logical errors; for correct implementation of its design; and for satisfaction of its requirements

2. Testing to verify the implementation of the design for one software element, e.g., a unit or module, or a collection of software elements

Source: IEEE

Synonymous with component testing.

Testing, usability—Tests designed to evaluate the machine/user interface. Determines if the communication devices are designed in a manner so that the information is displayed in an understandable fashion, enabling the operator to correctly interact with the system.

Testing, valid case—A testing technique using valid (normal or expected) input values or conditions

See Equivalence class partitioning.

Testing, volume—Testing designed to challenge the ability of a system to manage the maximum amount of data over a period of time. This type of testing also evaluates the ability of a system to handle overload situations in an orderly fashion.

Testing, worst case—Testing that encompasses upper and lower limits and circumstances that pose the greatest chance of finding errors. Synonymous with most appropriate challenge conditions.

See Testing, boundary value; Testing, invalid case; Testing, special case; Testing, stress and Testing, volume.

Touch screen—A touch-sensitive display screen that uses a clear panel over or on the screen surface. The panel is a matrix of cells, an input device, that transmits pressure information to the software.

Traceability—1. The degree to which a relationship can be established between two or more products of the development process, especially products having a predecessor-successor or master-subordinate relationship to one another, e.g., the degree to which the requirements and design of a given software component match See Consistency.

2. The degree to which each element in a software development product establishes its reason for existing, e.g., the degree to which each element in a bubble chart references the requirement that it satisfies

See Traceability analysis and Traceability matrix.

Traceability analysis—The tracing of:

1. Software requirements specifications requirements to system requirements in concept documentation
2. Software design descriptions to software requirements specifications and software requirements specifications to software design descriptions
3. Source code to corresponding design specifications and design specifications to source code

Analyze identified relationships for correctness, consistency, completeness and accuracy.

See: Traceability and Traceability matrix.

Traceability matrix—A matrix that records the relationship between two or more products, e.g., a matrix that records the relationship between the requirements and the design of a given software component

See Traceability and Traceability analysis.

Transaction (IT)—1. A command, message or input record that explicitly or implicitly calls for a processing action, such as updating a file

2. An exchange between an end user and an interactive system

3. In a database management system, a unit of processing activity that accomplishes a specific purpose, such as a retrieval, an update, a modification or a

deletion of one or more data elements of a storage structure

Transaction analysis—A structured software design technique, deriving the structure of a system from analyzing the transactions that the system is required to process

Translation—Converting from one language form to another

Source: NIST

See Assembling, Compilation and Interpret.

Trojan horse—Purposefully hidden malicious or damaging code within an authorized computer program

U

Unambiguous—1. Not having two or more possible meanings

2. Not susceptible to different interpretations

3. Not obscure and not vague

4. Clear, definite and certain

Unit—1. A separately testable element specified in the design of a computer software element

2. A logically separable part of a computer program

Synonymous with component and module.

UNIX—A multitasking, multiple-user (time-sharing) operating system developed at Bell Labs to create a favorable environment for programming research and development

Usability—The ease with which a user can learn to operate, prepare inputs for and interpret outputs of a system or component

User—Any person, organization or functional unit that uses the services of an information processing system

See End user.

User's guide—Documentation that describes how to use a functional unit and may include a description of the rights and responsibilities of the user, the owner and the supplier of the unit

Synonymous with user manual and operator manual.

Utility program—A computer program in general support of the processes of a computer, e.g., a diagnostic program, a trace program and a sort program

Utility software—Computer programs or routines designed to perform some general support function required by other application software, the operating system or the system users. They perform general functions, such as formatting electronic media, making copies of files or deleting files.

V

V&V—Verification and validation

Valid input—Test data that lie within the domain of the function that the program represents

Validate—To prove to be valid

Validation—Establishing documented evidence that provides a high degree of assurance that a specific process will consistently produce a product meeting its predetermined specifications and quality attributes

Validation, process—Establishing documented evidence that provides a high degree of assurance that a specific process will consistently produce a product meeting its predetermined specifications and quality characteristics

Source: FDA

Validation, software—Determination of the correctness of the final program or software produced from a development project, with respect to the user needs and requirements. Validation is usually accomplished by verifying each stage of the software development life cycle.

See Verification, software.

Validation, verification and testing—Used as an entity to define a procedure of review, analysis and testing throughout the software life cycle to discover errors, determine functionality and ensure the production of quality software

Source: NIST

Variable—A name, label, quantity or data item whose value may be changed many times during processing. Contrasts with constant.

Variable trace—A record of the name and values of variables accessed or changed during the execution of a computer program. Synonymous with data-flow trace, data trace and value trace.

Source: IEEE

See Execution trace, Retrospective trace, Subroutine trace and Symbolic trace.

Vendor—A person or an organization that provides software, and/or hardware, and/or firmware and/or documentation to the user for a fee, or in exchange for services, e.g. a medical device manufacturer

Verifiable—Can be proved or confirmed by examination or investigation

See Measurable.

Verification, software—In general, the demonstration of consistency, completeness and correctness of the software at each stage and between each stage of the development life cycle

Source: NBS

See Validation, software.

Verify—1. To determine whether a transcription of data or other operation has been accomplished accurately

Source: ANSI

2. To check the results of data entry, e.g., keypunching

3. To prove to be true by demonstration

Version—An initial release or a complete rerelease of a software item or software element

See Release.

Version number—A unique identifier used to identify software items and the related software documentation that are subject to configuration control

Virus signature—The file of virus patterns that are compared with existing files to determine whether they are infected with a virus or worm

VMS—Virtual memory system

W

Waterfall model—A model of the software development process in which the constituent activities, typically a concept phase, requirements phase, design phase, implementation phase, test phase, installation and checkout phase, and operation and maintenance, are performed in that order, possibly with overlap but with little or no iteration. Contrasts with incremental development, rapid prototyping and spiral model.

Workaround—A sequence of actions the user should take to avoid a problem or system limitation until the computer program is changed. They may include manual procedures used in conjunction with the computer system.

Worm—A programmed network attack in which a self-replicating program does not attach itself to programs, but rather spreads independently of users' action

Page intentionally left blank

9781604207491